FATAL FLIGHTS OF THE RICH AND FAMOUS

FATAL FLIGHTS OF THE RICH AND FAMOUS

ALASTAIR GOODRUM

Pen & Sword

AVIATION

First published in Great Britain in 2026 by
PEN AND SWORD AVIATION
An imprint of
Pen & Sword Books Limited
Yorkshire – Philadelphia

Copyright © Alastair Goodrum, 2026

ISBN 978 1 03613 808 0

The right of Alastair Goodrum to be identified as Author
of this work has been asserted by him in accordance with the Copyright,
Designs and Patents Act 1988.

A CIP catalogue record for this book is available from the British Library.

Typeset in Ehrhardt 10/13 by
SJmagic DESIGN SERVICES, India.
Printed and bound in the UK by CPI Group (UK) Ltd, Croydon, CR0 4YY.

The Publisher's authorised representative in the EU for product safety is
Authorised Rep Compliance Ltd., Ground Floor, 71 Lower Baggot Street, Dublin
D02 P593, Ireland. www.arccompliance.com

For a complete list of Pen & Sword titles please contact
PEN & SWORD BOOKS LIMITED
George House, Units 12 & 13, Beevor Street, Off Pontefract Road,
Barnsley, South Yorkshire, S71 1HN, England
E-mail: enquiries@pen-and-sword.co.uk
Website: www.pen-and-sword.co.uk

or

PEN AND SWORD BOOKS
1950 Lawrence Rd, Havertown, PA 19083, USA
E-mail: uspen-and-sword@casematepublishers.com
Website: www.penandswordbooks.com

Contents

Acknowledgements

The author gratefully acknowledges the contribution of text and image received from Dr Anna Nyburg of Imperial College, London, who gave generous access to material in her book: Nyburg, Dr Anna, *The Clothes On Our Backs; How refugees from Nazism revitalised the British fashion trade*, (Valentine Mitchell, London, 2020), in particular Chapter 6: pages 85-93: *Refugee Stories: Hats off to Otto Lucas: Gay, glamorous and German.*

Acknowledgement is also gratefully given to the following persons and institutions that have provided images or information. Johan Visschedijk of 1000aircraftphotos. com for images from the Ed Coates, Walter van Tilborg and W.T. Larkins Memorial Collections and from his own excellent collection; Keith Allison; Michel Anciaux; Ed Balaun; The Robert S. Ball Aviation Collection (USA); John Bennett and abpic. co.uk; Bibliotheque Nationale de France; Alan Brown and airliners.net; Peter Brown and blackbusheairport.proboards.com/forum; The Finnish Heritage Agency, museovirasto.fi; Erik Frikke; Jeff Gilbert of JGphotographics.com (Australia); Chris Goss for generous access to his book: *Bloody Biscay*; Mike Henniger of aerialvisuals. ca; Tony Hisgett; Peter R. Keating; Don Ramey Logan; The US Library of Congress; Nasjonal Biblioteket: The National Library of Norway; Enrique Perrella and SVZM Spotters (Venezuela); Adrian Pingstone; Mindy Smith (USA) for providing the photo of Audie Murphy from her father's archive; Delphine Stevens, Roy Chadwick's daughter for the photo of her father; Paul Ward and coolantarctica.com; Wikimedia Commons. The author acknowledges credit for jacket images as follows: Top: Roald Amundsen: courtesy of National Library of Norway, ref: NPRA701, public domain. Bottom: DC-2s courtesy of Peter's blog, www.patmcast.blogspot.com_2012_01_ teneningen-2 CCA-Sharealike 3.0.

The author thanks Jonathan Wright, Harriet Fielding and the team at Pen & Sword Publishing for their help and encouragement to get this book off the runway.

Conversion Table

1 inch	25.400 millimetres
1 foot (12 inches)	0.305 metre
1 yard (3 feet)	0.914 metre
1 mile (1,760 yards)	1.604 kilometres
1 cubic-foot	0.283 cubic-metre
1 square-yard	0.836 square-metre
1 acre	0.405 hectare
1 pint	0.568 litre
1 gallon (8 pints)	4.546 litres
1 ounce	28.350 grammes
1 pound [lb] (16 ounces)	0.453 kilogram

To convert temperatures from Fahrenheit to Celsius: subtract 32, multiply by 5, divide by 9.

Freezing point = 32 degrees Fahrenheit; 0 degrees Celsius.

Introduction

Aviation history is full of evocative stories about the evolution of aeroplanes, flying and the perils of air travel and there are many ways of looking at these. The theme of this book is to recall some of those perils through the eyes of a selection of internationally-famous celebrity air travellers. What brings them all together here is that, as well as presenting these personalities, the stories showcase aeroplanes from the golden age of biplanes to biz-jets and airliners. They also illustrate the fallibility of people and technology, while giving a flavour of the social progress of air transport over the past 100 years. Sadly, the climax of these particular stories culminates in air crashes that took the lives of the celebrities involved. While the final selection of the stars might be open to debate, the breadth of celebrity representation in these stories is very wide, being drawn from the fields of aeronautics; cinema; exploration; fashion; music; politics and sport. Mysteries and myths have grown up around some of these incidents and while some of these can be debunked, others will pose unanswered questions. All, though, will demonstrate that Fame and Fortune alone are no protection from Fate.

Chapter 1

Sir John Alcock

In 1913, the vision of Lord Northcliffe, proprietor of the *Daily Mail* newspaper, in offering a prize of £10,000 (approximately equal to [≈] £960,000 in 2024)[1] to the first person to cross the Atlantic Ocean by air, was a worthy incentive for airmen to sit up and take notice. This prize was thrown open to pilots of any nationality and its rules stipulated that the flight should be made between any point in the USA, Canada or Newfoundland and any point in Great Britain or Ireland and could be made in either direction. The distance involved was 1,890 statute miles at the narrowest point and it also had to be completed in one 'hop' of not more than 72 hours in duration.

Only four years had elapsed since, in 1909, the Frenchman Louis Bleriot had won an earlier Northcliffe prize of £1,000 (≈ £96,000 in 2024) for the first flight across the English Channel. It speaks volumes about the fragile state of aviation when, in 1910, Lord Northcliffe also offered a prize of no less than £10,000 for the completion of a flight from London to Manchester – a mere 186 miles – within 24 hours and even with two re-fuelling stops allowed!

Aeroplane technology, however, had progressed little since that date and even by 1913, the prospect of an aeroplane crossing the mighty Atlantic presented a hugely daunting challenge. While the challenge itself was halted by the onset of the First World War, Lord Northcliffe renewed his prize at the war's end in November 1918, by which time the intrinsic value of the prize sum had been diminished by the economic effects of the war to ≈ £473,000 in 2024 terms. However, the state of aeroplane and engine design; the science of aeronautics and navigation and the art of pilotage had developed and improved beyond measure; to such an extent that flying the Atlantic was no longer a pipe-dream – it had become a probability. John Alcock's name is instantly associated with the Atlantic exploit and that, indeed, was his pinnacle, but he deserves to be remembered for contributing even more to the world of early aviation.

John William Alcock, known to his friends as 'Jack,' was born in 1892 in Seymour Grove, Stretford, Manchester. Basford House was home to the Nutthall family, to whom his father was a coachman in private-service. Educated at St Anne's Parish School and Manchester Central High School, John left school in 1908 to become apprenticed to the Empress Engineering Company on Stockport Road, Manchester, a firm that built and repaired cars and motor-cycles and later, aero-engines.

In 1911, Jack left Empress Engineering to work for Norman Crossland. During the following year, John worked on an aeroplane engine sent for repair by its owner, Maurice Ducrocq, a prominent French aviator who ran a flying school at Brooklands, London. Having impressed Ducrocq with his work on the engine, it greatly helped his case when John asked Ducrocq to take him on permanently as a mechanic. Ducrocq agreed and employed him from January 1911 to early 1913 as chief engineer and later chief engineer and pilot.

During 1912, Ducrocq encouraged John to learn to fly under the former's tuition and he turned out to be an able pupil, going solo after just two hours instruction. He was awarded Royal Aero Club certificate number 368 on 16 November 1912. From that date until the outbreak of the First World War in August 1914, John Alcock extended his flying prowess to include cross-country flights and competitive, public flying races held at both Brooklands and Hendon airfields. John became a very competent aero-engineer with Ducrocq and carried out work for A.V. Roe (Avro), assembling aeroplanes and doing test and delivery flights for that company, which was also based at Brooklands.

By 1914, when he joined the Royal Naval Air Service (RNAS), John Alcock was considered, by his contemporaries and the aviation press, to be one of the top ten pilots in the country.

In December 1916 he was promoted to Flight Lieutenant and posted to a squadron on the Romanian front. However, before he could join his squadron, fighting on that front stopped and he was posted to No.2 Wing RNAS on the island of Lemnos, in the Aegean Sea off Gallipoli, where he arrived early in 1917. Flight Lieutenant Alcock spent ten months operating from the islands of Imbros, Lemnos, Lesbos and Tenedos and the Salonika mainland, finding plenty of action against Turkish, Austro-Hungarian, Bulgarian and German forces, in the form of fighter and reconnaissance sorties, anti-U-boat patrols, ground-attack sorties and long-distance bomber operations.

There was great excitement at Mudros for the arrival of a single Handley Page (HP) O/100 twin-engine, long-range bomber, serial number 3124. It was flown out from England to impart an element of strategic bombing to the region and it fell to John Alcock, as the most experienced pilot, to undertake bombing operations in it. His experiences attacking Constantinople, and other major Turkish targets at night, would stand him in good stead after the war. His navigation had to be spot on, not only to reach the targets but to get home again, too. He hit a munitions factory and shipping in the harbour at Panderma, on the southern shore of the Sea of Marmara, on the night of 6/7 August and between that raid and 2 September, the O/100 was employed on anti-submarine patrols over the Aegean.

On the night of 2/3 September, John Alcock with second pilot/bomb-aimer Flight Lieutenant H.L. Gaskell and flight engineer, Warrant Officer Second Class S.J. Wise, braved searchlights and anti-aircraft fire to strike for the first time at the railway station and marshalling yards at Adrianople, near the Turkish border with Greece and Bulgaria. Both bombing runs were made at 500 feet to secure good results and the operation involved a round trip of 600 miles and eight hours in the air over difficult, mountainous terrain.

Modern fighters were like gold-dust in this forgotten corner, with the first of a small number of Sopwith Camel single-seat fighters only arriving in the Aegean theatre on 26 July 1917. The very next day Flight Lieutenant Alcock flew a sortie in the newly arrived fighter to intercept a German seaplane escorted by two German fighters. In the capable hands of John Alcock, the Camel showed its superiority and he shot down one of the escorts into the sea. He was awarded a Distinguished Service Cross (DSC) and although the citation states the award was for the action on 30 September, it is more likely to relate to his fighter combat successes and for carrying out daring operations in the Handley Page O/100.

Upon his return to Mudros, Alcock flew the Handley Page O/100 on an operation to bomb Constantinople. His aeroplane was hit by anti-aircraft fire and forced to turn back, he crash landed in the sea off Gallipoli where he and his crew were captured. The days as a prisoner of war passed slowly and gave John Alcock plenty of time to dream about and plan in his mind's eye, how he might attempt to fly across the Atlantic Ocean.

Flight Lieutenant John Alcock DSC was demobilised on 10 March 1919 and now directed his thoughts and efforts to the Atlantic challenge. The *Daily Mail* renewed its offer of a £10,000 prize and a further £1,000 was added by businessman Lawrence Phillips and 2,000 guineas (£2,100) by the Ardath Tobacco Company, making £13,100 in total. John was keen to have a go, so he contacted the Aviation Department of Vickers Ltd., builders of the excellent Vickers FB27 Mk IV *Vimy* aeroplane that Alcock believed was capable of making the flight. Because they lacked a pilot, Vickers had not submitted an entry yet, but Alcock convinced them that the flight was possible and that he was the man for the job. He was offered a job as a test pilot from 1 March 1919, at a salary of £600 a year (≈ £25,800 in 2024).

The *Vimy* was about two-thirds the dimensions of the HP O/100 that John had flown during the war, but Vickers' standard machine had a range of 2,440 miles and was powered by two of the proven and reliable Rolls-Royce Eagle VIII, 360hp engines. There was a need to crack on with readying the *Vimy*, since most of the other entrants had a head start on Vickers. The cockpit was modified so that pilot and navigator could sit side-by-side, the front skid-bar was removed and the bomb-bay modified to take extra fuel tanks. However, there was still the question of who would be the navigator? This was answered by a twist of fate that brought together John Alcock and Arthur Whitten Brown at the Vickers factory in Weybridge, where Brown was attending a job interview. During his meeting Brown mentioned his great interest in navigation and as a result was asked bluntly if he could navigate an aeroplane across the Atlantic – to which he replied: 'yes.' Brown was taken to meet Alcock; they bonded well together and the Vickers Atlantic team was formed.

In April 1919, after flight-testing a standard Vickers FB27A *Vimy* IV, it was fitted with extra fuel tanks, dismantled, crated up and ready for shipping out in company with the main ground party, on the SS *Glendevon* to St John's, Newfoundland.

On 4 May, Alcock, Brown and a small advance ground party, sailed to St John's on the RMS *Mauretania*. The race was on. Alcock's group reached St John's, Newfoundland on the 13th. The crated *Vimy* and its party arrived in St John's on 26 May. All four competing teams were at different stages of readiness but all were affected by two major factors. The first was the current and prospective state of the weather and the second was finding and hiring a suitable take-off ground. Both proved to be very difficult problems.

Alcock's despair of ever finding a take-off ground was resolved when Freddie Raynham sportingly allowed Alcock and the *Vimy* team to use his field at Quidi Vidi for assembly and flight testing, while he repaired his own aircraft for another attempt (also unsuccessful). Now they could un-crate the *Vimy* and begin the preparations for their own attempt. All it needed then was good weather. The *Vimy* was then flown a short hop across town to a larger, more suitable field, offered free of charge by local haulier, Charles Lester. It was alongside Munday's Pond out at Cornwall Heights and they called it Lester's Field in his honour.

On Saturday 14 June, following a few days of high winds, the weather conditions eased and predictions were as favourable as they were ever likely to be. Alcock decided it was time to go.

Fuelled-up with 865 gallons of petrol; navigation charts and instruments, food and drink aboard and dressed in warm flying suits, Alcock and Brown settled themselves in the cockpit. Brown clipped a marine sextant to the dashboard; an Appleyard course-and-distance calculator to the fuselage side and a drift indicator was stowed beneath his seat. On his lap he balanced a Baker Navigating machine (a type of astrograph) along with his chart board. Brown's navigation method would be 'dead-reckoning;' his main chart was a Mercator projection with summer circle routes marked on a transparent cover that could be moved over his map. It was a tight fit in the cockpit!

Pre-flight check done, the engines were started, warmed up and Alcock gave the signal for 'chocks-away!' To gain speed, Alcock held her down as long as he dare, then at 4.13 pm Greenwich Mean Time (GMT), the *Vimy's* wheels left the ground and she was on her way to Ireland – 1,890 miles away at its closest point. Just clearing a wood at the end of the runway, Alcock circled inland to gain altitude before the *Vimy* crossed the coast outbound at 4.28 pm GMT on the 14th.

Around 6.00 pm, while climbing towards the clouds above, there was a sudden loud noise from the starboard engine. One of the exhaust pipes had split away from the engine casing and was flapping noisily, before flying off in the slipstream. Now un-silenced, the noise from that engine exhaust was deafening and the two men were obliged to communicate by pencil notes scribbled on pieces of paper.

After three hours, they were routinely checking all the instruments, fuel consumption and Brown pumped fuel between the tanks. Estimating they had flown 450 miles, Brown was keen to get a sun-sight for an indication of drift. For an hour, John inched the *Vimy* upwards, until it reached the tops at around 6,000 feet and Brown saw the sun through a break in the cloud and managed to get a sun-sight. He confirmed they were only slightly off course and gave Alcock a course correction before they plunged back into clouds again.

They were in a noisy, flickering, very cold world. Alcock again eased the throttle open gently and eased the stick back, to try to climb above the cloud, so that Brown could get another fix. At 12.25 am, he calculated their position to be about half-way across and slightly south of intended track. They had covered 850 nautical miles at an average of 106 knots.

At 3.10 am the first sign of dawn appeared and they were heading slap-bang into a black wall of cumulo-nimbus cloud that lay across their course. Unable to take avoiding action, the *Vimy* plunged into the blackness, to be tossed around like a leaf; shaking and vibrating; lit up by flashes of lightning and peppered with hail, as Alcock wrestled with the controls to keep her steady – while having no external horizon to help him. The airspeed indicator-head froze up and with no speed measurement, the aeroplane stalled into a spiral dive that took all of Alcock's skill as a 'seat-of-the-pants' pilot to juggle ailerons, rudder and engine speed and recover – but not before they dropped like a stone to less than 100 feet. The *Vimy* emerged from the murk, in a very steep bank with the wings almost vertical. Seeing foaming waves almost above him, it was sheer instinct that made Alcock's hands and feet correct the crazy attitude. Applying full power, he clawed his way upward again – while having to turn nearly 180 degrees to regain his easterly course. It was a very close thing!

At 7.20 am at 11,000 feet, a glimpse of the sun gave Brown a fix that confirmed they were north of intended course but, better still, only 80 miles from Ireland. Brown gave Alcock a course correction for Galway and suggested they go lower.

At 8.15 am Alcock shook Brown's arm excitedly and pointed forward. It was land!

They crossed the coast of Ireland at 8.25 am, slightly north of Galway. With the cloud-base, down at 300 to 400 feet, stretching out unbroken into the distance and hills ahead with their tops in the cloud, Alcock decided that, even though they still had ten hours-worth of fuel left – more than enough to reach London – he would not risk a crash now by flying across Ireland in such poor conditions. They had done what they came to do and he would land here at a suitable place. Turning south, it was just minutes before the masts of Clifden wireless station came into view and confirmed their position.

The first non-stop trans-Atlantic flight ended when the *Vimy* carrying John Alcock and Arthur Whitten Brown landed (or rather, crash-landed) in Derrygimla Bog near Clifden at 8.40 am (GMT) on Sunday, 15 June 1919. When men from the nearby wireless station asked Alcock where they had come from, he is said to have replied: 'Yesterday, we were in America!' Later, many press interviewers quoted him as follows: '"When I was in America yesterday," said Captain Alcock on Sunday – and then he broke off with a gasp at the immensity of the thing that had just happened.'

Brown wrote later that the distance [they had flown] from coast to coast was 1680 nautical miles; equivalent to 1933 statute miles and they had flown it in 16 hours and 12 minutes at an average speed of 104 knots (119mph). The aeroplane was in the air for 16 hours and 28 minutes, allowing for circling over St John's and the final diversion to the landing site. John Alcock had flown virtually immobile, with his hands or hand and feet always gripping the controls, for sixteen hours without a break, nursing the course of the *Vimy* with great accuracy. Arthur Whitten Brown had used all his skill to plot their course and position with few external 'fixes,' doing it with great accuracy mainly by 'dead-reckoning' navigation, while monitoring the engine and fuel management instruments in fearful weather conditions. Not least, the Vickers *Vimy* and its Rolls-Royce engines had performed magnificently. It was a tremendous joint achievement.

Within twenty-four hours, these airmen were global celebrities, feted wherever they went. In London they received their prize cheques from Winston Churchill, secretary of state for war, who informed them they were to be knighted immediately. Alcock and Brown let it be known that they would give £2,000 to the Vickers and Rolls-Royce ground support team for their part in the success. The King invited them to Windsor Castle where he invested them each as a Knight Commander of Order of the British Empire (KBE).

In July 1919, Sir John Alcock KBE DSC went back to his flight-testing and air-delivery work at Vickers. His contract was due to expire at the end of the year and he was keen to open a motor-car dealership in London. The *Vimy* was recovered from the Irish bog, repaired, cleaned then on 15 December 1919 in their presence, was handed over for permanent display at the Science Museum, London, where it can be seen today.

John Alcock turned 27 years of age in November. Vickers asked him to represent the company at the first post-war International Aeronautical Exhibition to be held in Paris in December. Alcock could have made the trip by train and boat but felt it was an excellent opportunity to display the new Vickers *Viking* aeroplane. On 18 December he planned to fly the new *Viking I Amphibian* to Paris and land it on the river Seine, in the full glare of publicity outside the exhibition venue. Arrangements had also been made for him to meet the 'conqueror of the Channel': Louis Bleriot and he was keen to keep the appointment for both reasons.

On the morning of the 18th, it was a cold, blustery wind that drove rain across Brooklands aerodrome. The forecast for Northern France was for strong wind, then fog on the route to Paris – although the forecast was for clear weather over the Channel and over Paris itself. Despite the Vickers management team urging him not to fly, Jack brushed them aside and insisted he could handle the weather even though he was going to fly solo.

Approaching the Normandy coast in the vicinity of Varengeville, just west of Dieppe, after an hour into the flight he turned inland on a direct track towards Paris. The aeroplane was not equipped for 'blind' flying and so, when the forecast fog materialised in his path, as it grew thicker he was forced ever lower in an effort to keep the ground in sight. Alcock had the option of turning back, or pressing on – and he chose the latter. The aeroplane was seen at about 1.00 pm by Monsieur Pelletier, a farmer working near the hamlet of Dreule, an isolated rural community near Cottévrard, 25 miles north of Rouen, apparently being tossed about as if by a strong wind. Suddenly, he said, the pilot seemed to lose control of the machine and it crashed into a field. Later, it was deduced that, while now flying very low in poor visibility, Alcock saw he was heading for a clump of beech trees and in trying to avoid them, one of the aeroplane wings hit a tree. The wing was almost certainly damaged enough to destabilise the aeroplane – which is probably the stage at which Monsieur Pelletier saw it in the throes of crashing. People nearby heard a loud, unusual noise and after searching, found the badly smashed wreckage, with Jack Alcock lying mortally injured inside. As Monsieur Pelletier gently lifted him from the wreckage Alcock was heard to whisper what was thought to be: 'Thank you. Thank you very much.' These were the last words he uttered, as he fell into a coma soon after. Alcock was moved on a cart to the Pelletier house, where Doctor Grandsire from Bosc-le Hard attended but, with a fractured skull and internal injuries to his chest and stomach, the doctor could do little for him. It was at this point that Gendarme Rabatel arrived to search the patient for possessions and evidence of identification. An identity card revealed that the victim was none other than the famous Sir John Alcock. Local Mayor Roussel was informed and telephoned the Prefecture in Rouen, which in turn, informed No.6 British Military Hospital, which immediately despatched an ambulance. At 4.30 pm the ambulance arrived at the scene and still unconscious and breathing very shallowly, Alcock was driven to No.6 Hospital in Rouen, where, without regaining consciousness, he was pronounced dead shortly after arrival. Within days the British government organised the return of Sir John Alcock, without any public fuss, to his family in Manchester. His body then lay 'in state' in the church of Holy Innocents, Fallowfield from Christmas Day until taken on 27 December, by RAF vehicle, in a slow procession through the streets to his funeral at Manchester Cathedral. He was buried with full military honours in Manchester Southern cemetery, to the sound of the Last Post and the engines of two RAF aeroplanes flying overhead.

The conqueror of the North Atlantic had set the stage for the dawn of international air transport.

Chapter 2

Roald Amundsen

22 June 1928

In such an august company of men like Ernest Shackleton, Robert Peary, Robert Scott and Fridtjof Nansen, the Norwegian Roald Amundsen was truly a giant among polar explorers of the early twentieth century. Having been the first to reach the South Pole; the first to navigate the fabled North-West passage and the first to fly across the Arctic Ocean and North Pole, no-one could match the range of his achievements in both polar regions.

Born in Borge on 16 July 1872, Roald Engelbreght Gravning Amundsen was the youngest son of a sea-captain father who raised his family on the outskirts of Norway's capital, Christiania (now Oslo). Not particularly bright in school, from the age of fifteen, Roald directed his energy into reading accounts of the expeditions of Sir John Franklin, whose quest to find the fabled north-west passage, between Canada and the Arctic, caught the boy's imagination. Together with his even greater fascination with the Norwegian Fridtjof Nansen's crossing of Greenland and his epic three-year unsuccessful expedition to reach the North Pole, these intrepid adventurers provided the catalyst for his own future polar career.

Upon leaving school, to please his mother, Roald studied medicine, but the sea was in his blood and when she died in 1893, Roald turned to life as a sailor. By 1900 he had a ship-master's licence and would always prefer to be known thereafter as Captain Amundsen. He did not marry, but gained a reputation as something of 'a lady's man' and had several affairs with married women. In 1907 he began a relationship with Sigrid Castberg, the young, attractive but bored wife of a solicitor. Their liaison lasted until 1913 when he transferred his attentions to another. Roald embarked upon the longest and most intense of his affairs with Kristine Elisabeth Bennett. Kristine, was a beautiful Norwegian lady, married to a much older, rich English businessman. She lived in England and it is to be marvelled that their intermittent, sometimes stormy, long-distance relationship was able to survive for almost ten years. His attentions wandered again when he met Elizabeth Magids (Bess) in 1922, while *en route* from Nome to Seattle after his *Maud* expedition. Bess was the wife of the owner of a chain of trading-posts in Alaska and Amundsen wooed her for many years. Divorced in February 1928, at age 31, Elizabeth actually managed to pin him down and became engaged to be married to Roald – but the marriage was not to be.

It was in 1897, that Amundsen – seeking a voyage that would help him towards his master's 'ticket' – had his first taste of polar exploration. As first mate of the *Belgica*, he was part of a scientific expedition to Antarctica, mounted by the Belgian explorer Adrien De Gerlache. The ship was icebound from March 1898 to March 1899 and upon his return to Norway, Amundsen felt he had gained enough experience to plan his own expedition and fulfil his long-held ambition to reach the North Pole.

Financing his projects was a constant source of difficulty for him since, although not a poor man, he lacked the scale of personal funds to cover his expeditions by himself. Having to beg, plead with and persuade others to invest in his ventures, was therefore always a source of grief to him in circumstances where he really wished to exert total control.

With outside finance and many creditors, Roald mounted his first expedition to find the North-West Passage. His ship *Gjoa* sailed from Oslo on 16 June 1903 and by September, despite technical problems with the ship; the weather-, sea- and ice-states were favourable enough to place him near the Simpson Strait – within sight of completing the Passage. However, another objective of this expedition was about establishing the position of the North Magnetic Pole[1]. This quest required a much longer stay in those Arctic waters and to venture out closer to the pole itself, in order to make the necessary scientific observations. With this land-borne objective in mind, a sound anchorage was found for the *Gjoa* at King William Island and Amundsen and his team operated from that base for the next two years. A great deal of magnetic, climatological and oceanographic data was collected which he passed on to other scientists and the Norwegian government.

Amundsen and his colleague, meteorologist Peder Ristvedt, set out by dog-sled onto the Arctic ice of the Boothia Peninsula, to set up a series of stations so that magnetic observations could be taken. The outcome of this demanding work was to be able to confirm that the position of the Magnetic North Pole in 1904 was very close to the position when it was first discovered by the polar explorer James Clark Ross in 1831.

In addition to his scientific observations, Amundsen also studied the habits and lifestyle of Inuit visitors to his encampment. In particular, he became so impressed by the efficacy and superiority of Inuit use of animal skins for winter clothing and footwear, over the European use of wool and other fabrics, that he employed skins and native clothing techniques very effectively on his subsequent polar expeditions.

On 13 August 1905, Amundsen's *Gjoa* expedition completed its work and he now continued his voyage to complete the north-west passage. *Gjoa* finally sailed into Nome, Alaska on 31 August 1906. It was a triumphant end to a long and arduous three-year expedition and Roald Amundsen had written himself into the annals of polar exploration.

FATAL FLIGHTS OF THE RICH AND FAMOUS

1908 found him planning a new polar expedition to retrace Fridtjof Nansen's steps using Nansen's ship *Fram* to drift across the Arctic Ocean for a projected five-year voyage. One objective was to see if any land mass actually existed out there in the vast Arctic Ocean and Amundsen also nurtured private thoughts of actually reaching the North Pole itself. His plans came unstuck when he was unable to raise sufficient funds to finance it. The public seemed to prefer to support more dramatic aspects of polar exploration and Amundsen's funding sources dried up. In an effort to find that public acclaim and to utilise his own already meticulously planned and equipped *Fram* expedition, Amundsen secretly re-aligned his attention to the South Pole.

Sailing from Norway in June 1910, Amundsen was expected to round Cape Horn and set sail for Arctic waters. Instead, when *Fram* reached Madeira, he let it be known to the world that he was heading for the Ross Sea in Antarctica and would attempt to reach the South Pole.

Amundsen and his team: Olav Bjaaland; Hilmer Hanssen; Sverre Hassel and Oscar Wisting, reached the South Pole on 14 December 1911. Applying his vast experience, his success came down to his choice of route and his equipment. His men were expert skiers and he used polar-hardy dogs, light-weight sleds and animal-skin clothing; keeping his operation as simple but as flexible, as possible. His route, though extremely arduous in its early stages, took him on a steep, tortuous – but short – climb up a glacier to a plateau, but thereafter the terrain was undemanding of his dogs and sleds and he reached the Pole relatively easily. Amundsen set out with fifty-two dogs and once clear of the glacier, he had no compunction in killing many of them for food for his remaining dogs and his men, and laid down a meat supply in store depots for the return journey. Spending three days at the Pole, the team took many sextant readings in order to verify their position accurately. By the time Amundsen returned to his base camp, he had only eleven dogs left – but all his five companions. Amundsen's achievement was rightly lauded in Norway but, in England, he was less than welcome, for the British have always loved a sporting loser and it was Robert Falcon Scott who became their perceived hero of the South Pole.

Despite the feelings of the British, Amundsen was in great demand around the world and as a result of writing, lectures and personal appearances, his finances recovered greatly and he was able to pay off his creditors. Having made some sound stock-market investments, by 1916 Amundsen was able to finance a new purpose-built ice-exploration vessel which, when launched in 1917, was named *Maud*. It is also significant that, while in the USA in 1914, Roald bought a Maurice Farman aeroplane and after shipping it back to Norway, he learned to fly it, before donating the machine to the Norwegian government when war broke out. His interest in aviation would take centre stage later.

Amundsen's next major voyage sailed from Norway on 18 July 1918, with a nine-man team, bound for the arctic waters of the North-East Passage, to the north of Siberia, intent on drifting across the Arctic Ocean to the North Pole. Ice-bound off Cape Chelyuskin, it was a whole year before the ship moved east again. By September 1919, still unable to turn northwards to begin the arctic drift project, the *Maud* was again gripped by ice off Ayon Island, in the East Siberian Sea, still 600 miles from the Bering Strait. It was not until 25 July 1920, that *Maud* finally sailed in to Nome harbour, having successfully navigated the north-east passage – which had, however, already been conquered – and having produced much scientific data – but failing to achieve the intended drift-exploration across the Arctic Sea. This and a subsequent expedition drained his financial resources and he needed to do something about that.

Roald Amundsen, having found a taste for flying while on a lecture tour of the USA in 1913, upon his return to Norway – where there was little by way of civilian flying training schools – used his status to obtain permission to learn to fly by courtesy of the Norwegian Army Air Service. Commencing his training at Kjeller airfield, near Oslo, in March 1914 under the tuition of Captain Einar Sem-Jacobsen on a Maurice Farman MF7 *Longhorn* biplane Roald, after a mishap or two on the way, qualified in June 1914 for the first pilot licence to be issued to a Norwegian civilian.

Amundsen now directed his attention and efforts towards using aeroplanes to reach the targets that persistently eluded him by other means. Aerial polar expeditions were to become as much an obsession with him as his past efforts by sea and land. In 1918 the American Curtiss aircraft company lent Amundsen a small *Oriole* biplane, which was packed in a crate and loaded aboard *Maud*, with the intention of using it to conduct short-distance survey flights over the ice. It was named *Kristine*, in honour of Roald's mistress. He recruited a couple of pilots and a mechanic to his little team and *Maud* would be captained by Oscar Wisting who would, once more, try to drift towards the Pole, while Amundsen had a go by air. *Maud* became stuck in the ice off Wrangel Island, East Siberia – going nowhere again; the Curtiss made just two survey flights before it crashed and was written off.

Roald's next venture, in May 1922, involved him purchasing a single-engine Junkers Larsen JL-6 seaplane, with the intention of flying across the North Pole from Wainwright, Alaska to Spitsbergen. His first Junkers, which he bought in New York, was written off in a crash landing due to engine failure, while Amundsen and his co-pilot/mechanic Oskar Omdal were ferrying it from New York to Seattle. Fortunately, they were both unhurt, but now Roald had yet more expense trying to obtain a replacement aircraft. With the replacement aeroplane delivered more safely to Wainwright, Alaska, by freighter in the winter of 1922/23, his plans for a trans-polar flight were back on track. Named *Elizabeth*, after his latest mistress, his aircraft was assembled at Wainwright but flying conditions were poor, so Omdal looked after the aeroplane while Amundsen travelled overland to spend the winter in Nome. His biggest difficulty, as always, lay in raising finance and that problem now overwhelmed him to the extent that he had instruct his brother to sell some of his property in Norway when, in 1923, his creditors turned on him.

Roald returned overland to Wainwright in May 1923. Omdal, meanwhile, had put skis on the Junkers, but a test flight damaged the left landing gear and wing then, when similar mishaps occurred on other flights, it was found the aircraft was simply not built for the stress of ski-landings and the damage had made it structurally unsafe. Amundsen was obliged to call off his trans-polar attempt and in November he returned, dejected, to Norway to face his creditors. *Maud* eventually made it back to Nome in August 1925 and was sold off to the Hudson's Bay Company to help pay off some of Amundsen's debts.

With his financial and family affairs in a mess, in 1924 he embarked on a lecture tour of the USA to raise funds. His exploits and reputation, though, were fading from public memory and this tour was far from successful. However, his luck was about to change for the better. While in New York, Lincoln Ellsworth, the son of an American multi-millionaire coal-mine owner, called upon Amundsen. Ellsworth was interested in polar exploration and after much discussion, he offered to finance a trans-polar expeditionary flight from Spitsbergen. They bought two twin-engine Dornier Do J '*Wal*' ('*Whale*') flying boats, serial numbers N-24 and N-25, in which to make the flight and a base was set up at Kings Bay (Ny-Ålesund), on Spitsbergen, the largest island in the Svalbard archipelago. Located 300 miles north of Tromsø, Norway and 700 miles from the North Pole, the main objective was to fly to the Pole, but then to return to Svalbard by a different route to undertake a geographical survey of the area between Spitsbergen and the Pole. The aerial expedition took off from Kings Bay at 5 pm on 21 May 1925. In N-24 was: Lief Dietrichson, pilot; Lincoln Ellsworth, navigator; Oskar Omdal, mechanic and in N-25: Hjalmar Riiser-Larsen, pilot; Roald Amundsen, navigator and Ludwig Feucht, mechanic.

Heading towards the North Pole, the aircraft soon encountered fog banks which, when left behind, revealed a seemingly un-ending expanse of featureless arctic ice. After eight hours in the air, in the morning of 22 May some open water was seen and when one of N-25's engines failed, it was decided to land on the sea, effect repairs and calculate their position. N-25 landed but, when N-24 followed her down, it sustained damage beyond the capabilities of the crew to repair, so they abandoned her. Although the two crews had landed less than half-a-mile apart, the drift-ice between them broke up and gave Ellsworth's crew, laden with such stores as they could carry on their backs, a treacherous journey to reach Amundsen's N-25.

Amundsen calculated they had reached position 87 degrees 43 minutes North (about 150 miles from the Pole) and it was clear that they would go no further towards the Pole – and would be fortunate if they made it back to Svalbard, too. But, if nothing else, they were six determined and resourceful men. To avoid the ice crushing N-25, they exerted an almost super-human effort and managed to haul N-25 from the water, up onto firm ice. Then they set about creating a runway on the ice. The profile of the Dornier's keel was such that it could make a take-off from an ice surface by sliding along under engine power, but that surface had to be as flat as possible. It took the party three-and a-half weeks of backbreaking work in those harsh arctic conditions, working long hours on just 1lb of food per day, to shift snow and ice and level the take-off run. On several occasions they almost lost the Dornier when the ice gave way and they had to drag it to safety, but their lives were at stake and self-preservation drove them on relentlessly. Finally, on 6 June, they found an ice floe that stayed intact and spent the next ten days preparing yet another ice-runway. On 15 June, stripping out all unnecessary equipment and with just eight hours of fuel left in the tanks, the six weary men boarded N-25. With Riiser-Larsen at the controls, N-25 was coaxed to slither off the ice and into the air. Eight hours later they landed on the sea off the north coast of Spitsbergen, where they were picked up safe and sound by a fishing boat, which took them to a rapturous welcome in Kings Bay. The party returned to Oslo where Amundsen was feted as the great national hero once again. However, in Amundsen's mind there was still one ambition he had, as yet, not fulfilled: to cross the Arctic Ocean by air, via the North Pole.

Roald opted for an airship solution next time. Lincoln Ellsworth came up with the funds and the Italian airship designer and adventurer, Umberto Nobile, provided the vehicle: the dirigible airship N-1, christened as the *Norge*. Amundsen and Riiser-Larsen met Nobile in Oslo; agreed the purchase and employed Nobile as captain of the airship for the expedition. On 7 May 1926, the *Norge* reached Kings Bay, where a hangar had been built for her and from where she would begin the voyage to the Pole. Some of Amundsen's thunder was stolen when the American, Lieutenant Commander Richard E. Byrd, US Navy, arrived in King's Bay with a Fokker Trimotor aeroplane, to attempt his own air assault on the North Pole – which allegedly he achieved on 9 May, with a 15-hour round-trip from Kings Bay. Amundsen was not at all deterred by Byrd's success and continued to prepare for his airship departure on 11 May.

In modern times, there has been some controversy about Byrd's claim to have reached the North Pole. His diary, unearthed in 1996, is alleged to state that, due to a serious oil-leak, he was obliged to turn the Trimotor back to Kings Bay while still an estimated 150 miles from the Pole. If this is true, then Roald Amundsen and his party aboard *Norge*, actually were the first to reach the North Pole by air.

With sixteen people on board (including Oscar Wisting; Lincoln Ellsworth and Umberto Nobile), *Norge* lifted off from Kings Bay and set course for Point Barrow in Alaska, via the North Pole. The Pole was reached – at last – on 12 May and flags of Norway, Italy and the USA were dropped as they circled that elusive goal. The onward flight to Point Barrow took them across the greatest tract of unexplored Arctic territory and in the early hours of 13 May, after 46 hours in the air, land-fall was made near Point Barrow. Despite the crew's physical exhaustion, the airship flight continued to Teller in Alaska, where *Norge* finally landed on 14 May after 72 hours aloft.

In 1928, Nobile, with the support of Mussolini's government, mounted his own air expedition to the North Pole in an effort to restore what he saw as his and his country's honour. He used a new airship, powered by three engines, named *Italia* and operated from the established base at Kings Bay/Ny-Ålesund on Svalbard.

Nobile made three voyages out from King's Bay. His third voyage, setting out from Kings Bay on 23 May 1928, would precipitate a chain of events that brought Amundsen and Nobile together in a way neither man could possibly have anticipated – nor wished.

During the morning of 25 May, the elevator controls malfunctioned, causing the airship to rise and fall alarmingly. Attitude and altitude could not be controlled and at 10.30 am on 25 May, *Italia* crashed at some speed onto the ice, off the Svalbard coast some 240 miles north-east from Kings Bay. The ventral control-gondola disintegrated and ten men inside, including Nobile, were thrown out, fortunately together with some stores, onto the ice below. Damaged, but duly lightened, *Italia* rose again, carrying off six crew working inside the envelope and engine cupolas – never to be seen again.

Miraculously, in addition to provisions and a tent, a field radio also survived and proved to be their salvation. Not until 3 June was a distress message picked up, purely by chance, by a Russian radio 'ham', but his alarm call initiated the largest air- and sea-rescue ever attempted up to that time.

Among those involved in the rescue effort was Roald Amundsen, who was in Oslo awaiting the arrival of Elisabeth Magids, his bride-to-be. Hearing of the plight of the *Italia* while at a formal dinner, he declared that he would offer his services to the Norwegian part of the rescue operation. Feeling he had much experience to offer, Roald sought an aeroplane in which he could mount a private search but, turned down – by Lincoln Ellsworth, for example – by 13 June he could find no financial backers. This state of affairs came to the attention of a Norwegian businessman in Paris, Fredrik Petersen. Petersen was aware that a prototype Latham 47 flying-boat, belonging to the French Naval Air Arm, was about to undertake a trans-Atlantic proving flight. He made representations on behalf of Amundsen, to have it diverted temporarily to the *Italia* rescue effort and on 15 June, the French government ordered the crew of four French Navy personnel to fly Latham 47, '02,' from its base at Caudebec-en-Caux in Normandy to Bergen in Norway. Amundsen travelled to Bergen, where he was to meet up with the flying-boat. He was accompanied by his friends Dietrichson and Wisting who were, of course, well-versed in polar navigation and Arctic conditions. Amundsen and Dietrich joined the crew of the Latham and on 17 June, flew north to Tromsø to refuel and take on stores.

At Tromsø there were already several flying boats being prepared for the search but Amundsen was impatient and did not wish to wait until the following day to join a combined air search. As a result, at 4.00 pm on 18 June 1928, Roald Amundsen and Leif Dietrichson, together with the French crew: Capitan de corvette René Guilbaud, pilot; Emile Valette, radio; Lieutenant Albert de Cuverville, navigator and Gilbert Brazy, mechanic, took off in the heavily-laden flying-boat and set course for Kings Bay on Svalbard, about 650 miles north across the Barents Sea.

They were never seen again.

The first clue to the fate of their mission came on 13 August, when a damaged stabilising float, identified as from the Latham, was found in the sea near the Norwegian island of Fugløya (near Bodø). Then, on 13 October, a fuel tank, also identified as from the Latham, was found on a Norwegian beach, followed by the discovery of a second fuel tank a few days later. These were the only items of wreckage ever found.

Chapter 3

Juan de la Cierva

9 December 1936

It is one of life's ironies that Juan de la Cierva, the man who invented possibly the safest aeroplane to fly, met his premature death in an airliner crash. Cierva was a Spanish aeronautical engineer, pilot and most famous as the inventor of the gyroplane as a practical, airworthy type of flying machine, generally accepted as the forerunner of the helicopter.

Born on 21 September 1895, Juan de la Cierva y Codorníu was the eldest of two sons of a wealthy Murcian lawyer – also named Juan – who, at various times, held a number of important ministerial posts in the Spanish government. When the family moved to Madrid in 1904, Juan junior began to show a great interest in the new science of aeronautics. Reading anything he could lay his hands on, he devoured aeronautical information and took every opportunity to watch the antics of famous pilots of the day, when they visited Madrid to exhibit their flying skills in their oh-so fragile aeroplanes.

Attending the Escuela Especial de Ingenieros de Caminos, Canales y Puertos in Madrid, Juan Cierva studied civil engineering and theoretical aerodynamics for six years between 1913 and 1919. After his graduation, having acquired a great deal of practical and theoretical knowledge of aeronautics, he never actually practised as a civil engineer, preferring to pursue a career as a scientist in aviation. In addition to his graduation, it was also in 1919 that he married Maria Luisa Gomez-Acebo, whom he first met while at university. Furthermore, with the help and influence of his father, that year saw Juan junior move into Murcian regional politics, in which he – not entirely enthusiastically – continued to work until 1923. Cierva had read of the ideas and efforts – un-successful in practise – of several aero-engineers around the world on the subject of 'auto-rotation' and he became particularly interested in utilising freely-rotating wing-blades to provide lift and stability, in a concept that he initially called a 'gyroplane,' later generally referred to commercially as an 'autogyro.' The scientific pursuit of this dream was to become an obsession for the rest of his life.

Juan wanted to find out if a set of wings, or blades – mounted atop a pylon attached to an aeroplane fuselage and rotating like a horizontal windmill – would also generate lift. The disc of three, four or five blades – each shaped with an aerofoil cross-section – would be drawn through the air and thus rotate as a disc, by the forward thrust from a separate engine-driven propeller mounted conventionally in the fuselage. With financial backing from his patents Juan, working in Spain, set out to prove his theories.

By experimenting with models, it became clear to him that the instability – and crashes during the take-off runs – was caused by both the overall gyroscopic force of the spinning rotor and an unequal lift from each rotor blade as it went round the circle. The blades on his small models were more flexible at their roots and this seemed to make the whole machine much more stable when he tossed them from the roof-tops. When a rigid blade advances, it generates lift, but as it retreats on the other side of the circle it loses lift and this asymmetric pattern is the primary reason for instability and its reluctance to fly. This latest discovery enabled Cierva to make the most significant of his early modifications – one that was to have huge beneficial implications for the future of rotary-wing flight. In his *C.4* model, gone were the rigid blades; now replaced by joints at the hub that hinged horizontally and vertically, allowing free movement as they went round the circle. Not only did this resolve the asymmetric-lift issue but it also counter-acted the gyroscopic effect problem. Now his invention had lift-off!

The first successful flight of the *C.4* was made on 17 January 1923 and it was only a few days later that the low-speed safety of the autogyro design was amply demonstrated. Following an engine failure in flight, although the forward speed rapidly decreased, the pilot landed the machine under auto-rotation from a height of about fifty feet. The safety factor, inherent in this design, was proven.

The next major task was to develop the short take-off potential of the autogyro and then undertake flights over greater distances than those made to date. Short take-off was all about the rotor being spun up to a speed that produced lift – preferably before the ground run began. If the forward engine was relied upon, to move the aeroplane along the ground until that motion gradually spun the rotor up to a point where the rotor lift took over and the aircraft left the ground, then that process would take time and distance to achieve. However, if the rotor could be 'pre-spun' up to – or even exceed – normal lift revolutions per minute (rpm), then the take-off run would be very considerably shortened. As soon as the brakes were released, the aeroplane would virtually 'hop' into the air and climb away as additional momentum and airflow, generated by the effect of the forward engine, caused the rotor to gather more speed.

First tested on his *C.11*, it was in 1930 that he linked the free-wheeling rotor by a drive shaft and clutch mechanism to the forward engine. The arrangement proved unreliable and too heavy for the *C.11* to lift but, as an effective alternative, Juan found that by re-designing the tail structure, on the ground the elevators could be locked in a near-vertical position that deflected the airflow from the front engine, up through the rotor disc to get it spinning. This airflow-deflection innovation, perfected in the

C.19 Mk II, was a great improvement and one that enhanced the independence of the aeroplane; having an immediate effect on its marketability.

Having taken out many worldwide patents on his designs and innovations, Juan was by now famous across the globe, both in the aeronautical world and to the general public, since his revolutionary flying machine was guaranteed to generate media interest wherever it was seen. Always seeking more finance to continue to improve the performance and reliability of his designs, when investment was not forthcoming in Spain, in 1925 Juan Cierva was persuaded to relocate to England. Following a successful demonstration of his autogyro at Farnborough, the British government injected some funds and when he received more private finance from a Scottish industrialist James G. Weir, Juan was encouraged to set up the Cierva Autogiro (note the different spelling) Company Ltd., with Weir as its chairman. Juan lived in England and it was here that he not only continued to refine his rotor designs but also used England as a base from which he travelled the world, monitoring the construction of autogyros in those countries that had bought licences to manufacture his designs; companies in France, Germany, Japan, Russia, Spain and the USA being the main licensee locations.

One of the difficulties Juan Cierva had encountered up to this time was that he was not a pilot himself and had to rely on the reports of others and his own observations from the ground. In 1926 he decided to do something about this and learned to fly while working at the Avro factory in Hamble, qualifying for his fixed-wing private pilot licence on 27 January 1927. The famous Australian pilot, H.J. 'Bert' Hinkler, who was a test pilot at Hamble, then taught him how to fly an autogyro.

At the invitation of the French Minister of War, on 18 September 1928, Juan made the first international flight by an autogyro from London to Paris, for another major sales initiative. Among those interested in purchasing manufacturing rights for the gyroplane, was Harold Pitcairn, an American aeroplane designer. By March 1932, in collaboration with Pitcairn in the USA, Juan overcame the difficulties he had encountered with the direct-drive, when a lightweight drive and clutch, designed by Pitcairn further enhanced the spin-up system and in 1933 was installed in his *C.19 Mk IV* aeroplane. Thus, the *C.19* model represented another important milestone. In this version, the autogyro could still retain its free rotor – running very smoothly now on ball-bearing joints – but it could be temporarily spun up to high revolutions by the drive from the forward engine while the machine stood at rest on the ground. The direct-drive mechanism was not needed during the landing phase.

One of the final key developments for Cierva's autogyro designs was to add a mechanism that allowed the rotor disc to be tilted in any direction by the pilot, giving him/her control of direction by movement of a control column attached to a tilting rotor hub. This development, another highly significant innovation, dispensed with the need for separate ailerons and stubby wings – and was the final piece of the puzzle that gave the autogyro its ultra-short take-off; its short or almost vertical-landing capability and confirmed its amazing ability not to stall at ridiculously low speeds. This brought about the Cierva *C.30* model, the prototype of which, G-ACFI, flown by Juan, was demonstrated to the public at Hanworth aerodrome on 27 April 1933.

By 1936 Cierva had finally perfected what he called an 'autodynamic' rotor hub. By over-rotating the blades and automatically manipulating their angles of attack, this system enabled almost vertical take-off from a standing start. Under direct-drive, when the rotor reached about 200rpm, the drive clutch was disengaged by the pilot and the autogyro fairly leapt about 20 feet into the air with no forward run. Once airborne, the forward engine and the rotor settled easily into their usual equilibrium and the autogyro performed as expected.

Juan de la Cierva's patient development work brought him just rewards and several accolades were bestowed on him. He received a Daniel Guggenheim Gold Medal (USA); the Royal Aeronautical Society of Great Britain awarded him a silver medal; the International Aeronautical Federation, its Air Gold Medal, all in 1932 and the US Franklin Institute awarded him the Elliott Cresson Gold medal for engineering in 1933. In 1934 – despite him being a monarchist at heart – the Spanish Republican government made him a Knight of the Order of the Republic.

Focussed on his work and travelling the world to further his business interests, saw Juan making many trips aboard airliners and it was this scenario that brought him to London's Croydon airport on the morning of Wednesday, 9 December 1936. He was flying to Berlin via Amsterdam on business and had booked a seat on a scheduled Royal Dutch Airlines (KLM) flight from Croydon. There was a thick fog blanketing the area, with visibility about 20 to 25 yards. It was, however, quite normal for airliners to take off even under such poor visibility conditions. In 1931, the airport had taken steps to lay out a straight, white chalk line on the grass surface, right across the airfield, indicating the general east/west direction of take-off. In times of poor visibility, a pilot could align his aircraft with this guide-line, keeping it in view from the cockpit even though he might not be able to see the perimeter of the airfield. It was particularly aligned so that aircraft could take off towards an area that was free of obstacles or high ground. A Swissair DC-2 had taken-off under these conditions some twenty-five minutes before the departure of Cierva's KLM flight and other departures had been made safely earlier that morning.

In command of the KLM DC-2 was Captain Ludwig Hautzmayer, an Austrian by birth who, at the age of 43, was an experienced airman and was a fighter pilot ace in the Austro-Hungarian Air Force during the First World War. He joined the Dutch national airline KLM in January 1936. His aeroplane today was an American-built, 14-passenger capacity, Douglas DC-2-115E airliner, registered in Holland as PH-AKL and bearing the name *Lijster (Thrush)* on its nose-cone.

At a little after 10.00 am, with thirteen passengers settled on board (see Appendix 1), Captain Hautzmayer started engines and cautiously followed a tractor, equipped with a bright red light, out to the start of the white line. It was approaching 10.30 am as he ran up the engines; released the brakes and began his take-off run towards the west, straining to keep the white line in view through his left cockpit window.

An hour or so earlier, there had been fourteen passengers in that cabin when, around 09.00, Captain Hautzmayer made his first attempt to depart. Louis Warshaw, the director of a timber company, was anxious to get to Danzig for urgent business reasons. Later that day, Mr Warshaw said he decided to leave the aeroplane and make alternative plans, telling a reporter:

> 'I ought to be dead. I can't believe I am alive. Providence must have warned me [as] I waited three times for take-off.'

In contrast to this lucky escape, is the story that Miss Hilde Bongertman, the 23-years-old stewardess, actually swapped places with the duty steward on this flight, in order to make a swift return to Holland.

In addition to Juan de la Cierva, there was another celebrity on board that fatal day: 74-years-old Rear Admiral Arvid Lindman, a retired Swedish politician. Lindman had a distinguished naval, industrial and political career, which saw him rise to become the Prime Minister of Sweden for two separate terms; 1906-1911 and 1928-1930.

It would be trying enough in clear weather for a pilot to control his aeroplane and keep the white line in view, but in a 30-yard pea-souper at up to 80mph, it was a difficult proposition. Wheel tracks in the grass showed that Captain Hautzmayer managed to hold on to his westerly run, parallel to the line, for only 200 yards before the aeroplane veered off to the left, crossing the line towards the south. By the time it reached the aerodrome perimeter, it was almost at right-angles to the white line and was barely airborne. One hundred yards from the perimeter, the port undercarriage wheel demolished a light wooden frame in its path. At eighty yards both wheels left the ground, but both of them hit the iron-railing fence of the airport boundary, knocking down a section, before uprooting a wire fence surrounding a tennis court located outside the aerodrome.

Heading in this direction at such a low height doomed the airliner to disaster. It was flying towards rising ground, but not gaining enough height fast enough to clear unseen houses on the slope towards which it was rushing headlong in the fog. In a matter of seconds, the DC-2 brought death and destruction to a quiet suburban avenue. Barely a mile from the start of its take-off run, Hillcrest Road, in the Beddington/Purley district, bore the brunt of the devastation and as its name suggests, the airliner had failed to gain the 150 feet or so, that would have seen it safe. Most of Croydon aerodrome lays within the 220 feet (70 metre) Ordnance Survey contour line, while Hillcrest Road straddles the 330 feet (100 metre) line and a stone's throw beyond that stands the top of Russell Hill at 360 feet (110 metres).

First, the airliner struck Nos. 12 and 14, a pair of semi-detached houses on the north side of the road. Spinster sisters, Gertrude and Maude Ault had a lucky escape as their house lost its chimney; falling through the roof but without injuring them. The stricken airliner hit the gable end of the second of the pair, owned by Mrs D. Vale. Its roof was demolished as shattered brickwork cascaded to the ground; spraying adjacent houses with shrapnel and debris. Just minutes before the airliner struck, Mrs Vale's small child had left the upstairs room in which she was playing. Cartwheeling across the road, the port wing of the aeroplane snapped-off a telegraph pole; ripping that wing from the fuselage, to lay in the road. The rest of the machine careered across the road, piling upside-down against the end of another pair of semi-detached houses on the south side; coming to rest against the front wall, tail in the air with the nose buried in the front garden of No.25. A deathly silence lasted just seconds before the remains of the aeroplane and house erupted into a blazing inferno. Flames leapt 40 feet into the air as a loud explosion ignited 500 gallons of petrol from ruptured fuel tanks. For Juan Cierva and most of those souls on board, the force of impact, fire and the tremendous saturation of the air by carbon monoxide from the explosion, snuffed out their lives in an instant.

No.25 bore the brunt of the damage. Both engines and propellers smashed through the walls before coming to rest in the centre of the almost entirely demolished house. Flames engulfed the house which, by a miracle, was unoccupied and had been so for a couple of months. Its previous owner, Mrs E. Faith, had handed over the keys to the new owner just days earlier. Incongruously, the For Let or Sale sign still stood intact by the gate. Next door, in No. 23, Miss Dora Kouvarine, a Russian governess, ran screaming from the house as the outside wall of the room she was in partially collapsed and flames licked through the opening. Fortunately, Mr Charles Koenig, her employer and his two children were out of the house at the time. She was taken in and comforted by Mrs Rivet at No.27.

Residents ran to the rescue but it was in vain. Fierce flames and heat kept them back. About to leave his home in Hillcrest Road, local GP, Doctor Lankester, decided that with the fog cutting visibility down to 10 to 15 yards, it was far too thick to risk using his car. Hearing the crash and seeing the flames, he ran blindly through the fog to the scene, but there was little that could be done for all but one of the broken bodies he found. To their credit, fire brigades from the surrounding district were quickly on the scene; facing a mammoth task to quell the petrol-fuelled inferno. It took three hours to bring it under control, during which time one fireman was injured and burned by falling wreckage.

Out of this carnage, though, staggered two dishevelled figures. Passenger, Walter Schubach, hands outstretched in front of him; badly burned about his arms and head, stumbled as far as the road then collapsed. It is not known how on earth he managed to get out of the wreckage. Aided by Dr Lankester, he was taken to hospital where he remained in a critical condition for several weeks. Stewardess, Hilde

Bongertman, clasping her hands to her head, ran screaming into the road. Somehow surviving the catastrophic impact, she had squeezed or been thrown through a split in the fuselage near a burst window. With flames crackling and smoke all around, she fell eight feet to the ground. Speaking later from her bed in Purley hospital; bruised; blood-stained; her hair singed; nursing two broken ribs and still dazed and trembling, Hilde said:

> 'After I climbed out, I was semi-conscious and in great pain. I have a vague recollection of groping about blindly and screaming until I was gathered-up and taken into a nearby house by a kindly person.'

Flight engineer, Pieter van Bemmel was flung into the garden and found showing some signs of life, but he died in an ambulance on the way to Purley hospital without regaining consciousness. These were the only survivors.

Press coverage of this disaster had a hard job vying for space with the announcement, in the House of Commons on 10 December, of the abdication of King Edward VIII, which inevitably grabbed the headlines. Apart from the names of those who died and an outline of events leading up to the dreadful scenes in Hillcrest Road, which also made the news on the 10th, nothing further was reported until an inquest on 16 December. This inquest was adjourned until 12 January 1937, by which time it was expected that the report of an Air Ministry accident investigation would be available.

In the circumstances, other than circumstantial evidence there was little the investigators could come up with. No-one capable of identifying or explaining the reasons for the crash survived. It was assumed, rather than proven, that there were no mechanical causes and so attention fell upon an interpretation of the pilot's assumed actions and judgement. Air Ministry investigator, Major J.P.C. Cooper, said visibility on the airfield was 30 yards at the time and made the point that the pilot ought to have closed the throttles and aborted the take-off run when he lost sight of the white line. It was considered he had adequate room to do this. At the point of impact, both engines were clearly giving power, but when the airliner strayed from the white line – an action considered unintentional – in an ever-increasing turn, the left wheel stayed in contact with the ground for twice the distance (450 yards) compared to the right wheel. The aeroplane was skidding or side-slipping across the grass and by the time it reached the perimeter of the airfield it could be said to be flying, but was only four- or five-feet off the ground. The gyro-compass and gyro-direction indicator were found to be in working order when removed from the wreckage.

From this description, it seems likely that, having become disorientated, in continuing the take-off and seeing houses in his path, the pilot may have tried to climb sharply with insufficient flying speed, causing the aircraft to stall.

The verdict of the jury at the resumed inquest was that the crash was an accident due to the pilot losing his direction owing to fog. The jury added a rider to its verdict:

> 'We are strongly of the opinion that the final decision as to whether aircraft should take off in conditions of low visibility should rest, not with individual pilots, but with an officially appointed authority of the airport.'

The Cierva Autogiro Company lost its charismatic, energetic leader and as a result, the progress of autogyros faltered. The ground-breaking work of Juan de la Cierva found its way into the development of the helicopter, the design of which had sufficient advantages to eclipse the autogyro in the future of aviation. Eclipse – but not eliminate, since in modern times there remains a thriving market for autogyro products in the leisure and scientific domains – in no small part encouraged by the antics of a Wallis WA-116, *Little Nellie* in the James Bond film, *You Only Live Twice*[1].

Chapter 4

Amelia Earhart

2 July 1937

By 1937, Amelia Earhart was one of the most famous aviators – male or female – in the world. That is both a fitting acknowledgement of her career achievements and also to her mysterious disappearance on her final flight – an enigma that still surrounds her name today. Noted for having a determined, independent disposition and being a firm advocate of women's rights and equality, her exploits in the air and on the ground, in the short period between 1921 when she learned to fly and 1937 when she mysteriously disappeared, made her a worldwide legend in her own lifetime.

Born in Atchison, Kansas, USA on 24 July 1897, Amelia Mary Earhart was the eldest of two daughters – the second being Grace Muriel (born 1899) – of Samuel 'Edwin' Stanton Earhart, an attorney and his wife Amelia Jane (neé Otis). Amelia and her sister grew up with their affluent maternal grandparents in Atchison, while their parents lived in Kansas City, visiting them during school holidays. The girls also had the benefit of a private education while living with their grandparents. By now, though, the marriage had become very strained and Edwin's wife decided to take the girls to live with friends in Chicago. Amelia graduated from High School in 1915 and attended Ogontz college in Philadelphia but did not finish her course. In 1917, during a visit to her sister, who was studying at a college in Toronto, Canada, Amelia volunteered as a Red Cross nursing aide at Spadina military hospital in the city.

In 1918, Amelia developed a sinus infection that required surgery and the condition troubled her for many years. For recuperation, she went to live with her mother in Northampton, Massachusetts and a year later enrolled on a medical course at Columbia University in New York. By the start of 1920 her parents were reconciled and living together in Los Angeles, California, where Edwin, having managed his alcohol addiction, was practising law. In an effort to support her mother in particular, Amelia dropped out of her medical course and moved back into the family home in Los Angeles, taking on a variety of jobs to earn money.

In 1921, Kinner Field, owned by the local municipality was also home to a one-person, one-aeroplane, flying school run by 24-years-old Mary Anita 'Neta' Snook, one of the earliest of American female pilots. It was here, on 3 January 1921, that

"

Amelia Earhart took her first flying lesson with Neta Snook in a Curtiss J-4 *Canuck*. After six months she showed signs of becoming a competent pilot but, in Neta's opinion, she would never be much more than that. History proved her right, too, since Amelia, along the way, did acquire a reputation for her shaky landings – and the occasional shaky take-off!

Her first solo must have been a little before 15 December 1921 because it is believed that, on that date, she passed her test for a US National Aeronautics Association (NAA) licence. Becoming disenchanted with the lumbering Curtiss J-4, Amelia worked hard to accumulate enough money to buy the prototype of Bert Kinner's own-design aeroplane, called a Kinner *Airster*.

After several heavy landings, came Amelia's first crash. With Neta in the front seat and Amelia in the back doing the flying, they had flown over to Goodyear Field, about 6 miles from Kinner, to look at a new aeroplane. On taking-off for the return to Kinner field, the engine was not pulling well and the *Airster* could not clear some trees. It stalled and crashed, breaking the propeller and undercarriage. Amelia and Neta emerged none the worse for their fall, but the poor *Airster* needed Bert Kinner's loving attention to get it flying again.

In May 1923, she felt it was time to try for her official FAI flying certificate. Possession of an FAI certificate allowed her to make attempts on any official world record and most importantly, ensure that the international governing body would recognise and ratify such records. The following month, Amelia sold her Kinner *Airster* and bought a car. The sale may have something to do with a decline once again in the family finances that culminated in the divorce of her parents in May 1924, when her mother moved to Medford, Boston, Mass., and Amelia and her sister moved in with her.

From 1925 and over the following few years, there came a new strand to Amelia's adult career. She took a post as a full-time, live-in social worker and active advocate of opportunities for women, on the staff of Denison House Settlement in Tyler Street, Boston. Although not formally trained in social work, she pitched herself whole-heartedly into this sheltered home, run by open-minded, liberal, middle- and upper-class, often college-educated women. Amelia worked tirelessly for the Denison Settlement cause both locally and at national level, gaining a reputation as one of the most promising social workers of her generation – an aspect of Amelia's life that is often overlooked.

One day in April 1928 while working a shift in Denison House, Amelia took a telephone call from Captain Hilton Howell Railey a public relations man; friend of and 'fixer' for the wealthy American publisher George Palmer Putnam. Right out of the blue, he asked Amelia if she would like to fly the Atlantic Ocean. Like a shot, she said 'Yes!' and the die was cast. Her life would change from social worker to world-famous aviator. Railey, living near Boston, had Amelia's name and reputation as the best-known female pilot in the city, drawn to his attention by some Boston acquaintances. Mrs Frederick E. (Amy Phipps) Guest, an extremely wealthy New England socialite, financed the purchase of a Fokker VII and covered the costs of an expedition for two male pilots: Wilmer 'Bill' Stultz and Louis E. 'Slim' Gordon and Amelia to fly from the US to Europe.

The Fokker F VII Tri-motor aeroplane, registration NX4204, was originally bought by Commander Richard E. Byrd to fly across the Antarctic. Naming it *Friendship*, and fitted with a water-float undercarriage, it was flown to Trepassey in Newfoundland for the trans-Atlantic flight. Amelia would receive no payment for her part in the venture – although both her companions were well-paid – and would do none of the flying. With Amelia sitting uncomfortably in the stripped-down cabin, the Fokker Trimotor set out from Trepassey on the morning of 17 June 1928. Just under twenty-one hours later, tossed around in the sky *en route*; with a misfiring left engine; running low on petrol and unsure of their position, the Fokker landed on the sea off what turned out to be Burry Port, in south Wales. One wonders about the value to the furtherance of aviation by Amelia's participation in such a venture. At first sight, it does not seem to have been a good deal for Amelia or her quest for female equality but, since she always craved the chance to be some sort of pioneer in aviation, she seemed content to sit in the back cabin: 'I was just baggage; like a sack of potatoes,' she said, '[but] when a great adventure is offered you, you don't refuse it.'

Back in the US, there were ticker-tape welcome parades in New York and Boston; meetings with the great and good – including President Calvin Coolidge at the White House and a book and publicity tour deal with George Putnam's organisation. Amelia spent that summer as house guest with Putnam and his wife Dorothy at their magnificent estate in Rye, a coastal suburb of New York. It was here that she wrote her first book, which was published within the year[1]. Amelia Earhart was now a global commodity and George Putnam had set his sights on reaping the benefit of his considerable investment in her. It would soon become apparent, too, that his interest in Amelia went beyond a purely business relationship.

Amelia flew an Avro *Avian*, with its English registration and a US tail-mark '7083', on her own first solo long-distance flight in September 1928. Privately disillusioned with the arrangements for the Atlantic flight, she was keen to establish her own public credentials as a pioneering pilot. This project with the *Avian* was the first flight made by a female pilot across the US from Boston on the east coast to Los Angeles on the west coast and back (about 5,500 miles round trip).

Having sold her Avro *Avian* in July 1929 Amelia acquired a second-hand Lockheed *Vega-1*, NC6911, single-engine, high-wing monoplane with a four-seat cabin, plus the pilot. She tried her hand in the 1929 National Air Races which were held for the first time at Cleveland Municipal airport, Ohio. This was dubbed the 'Powder-Puff Derby' and with a prize fund of $25,000 up for grabs, Amelia declared her Vega NC6911 as too unreliable so Lockheed kindly and quickly, traded it for a new Vega-1, registration NC31E. Sixteen competitors finished the gruelling race, with Amelia placed third (prize of $875; ≈ $20,000 in 2024[2]) of the ten 'heavy' class aircraft, behind winner Louise Thaden and second-placed Gladys O'Donnell.

As a result of spending so much time in each other's company during the race, several of the leading US female pilots discussed forming an organisation to further the cause of women pilots – of whom there were just 117 licenced in the USA. At a meeting at Curtiss Field, Long Island, New York on 2 November 1929, the 'Ninety-Nines' (named after the number of founding members) was founded and is still going strong to this day. Amelia Earhart was one of those founding members and in 1931 she was elected as its first president.

By June 1930, though, she was perhaps really only tinkering round the edges of aviation, for example setting relatively minor women's speed records under different weight parameters in a borrowed Lockheed *Vega* DL-1, NC497H at Grosse Isle airport, Detroit and being dragged all over the country to such an extent that she seems almost to have lost her sense of personal direction. What Amelia really needed was a major flying project to keep her at the forefront of the world's public and media and give her a real purpose.

She was also drawn into an ever-deeper relationship with George Putnam, through whom she was constantly involved in an endless round of fee-generating enterprises. In mid-September 1929, Dorothy Putnam took steps that eventually led to her divorce from her first (of four) husband of eighteen years before the end of the year. She left the marital home for Reno, Nevada in order to begin the divorce process, leaving the field open for Amelia and George to take their relationship to a new level. After much soul-searching and Amelia drawing up what, these days, would be called a pre-nuptial agreement, the 'promotee' finally married the 'promoter' when Amelia tied the knot with George Putnam, in a quiet civil ceremony on 7 February 1931 – the second of his own four marriages.

Amelia still had her sights set on a solo crossing of the Atlantic and in early-1932, she and George Putnam began to plan such a flight. She bought an ex-demonstrator Lockheed *Vega* 5, registration NC7952, on 17 March 1930, which she damaged nosing-over after a bad landing at Langley Field, Virginia on 25 August 1930. Now, a couple of years on, her engineering technical advisor, the highly-respected Norwegian Bernt Balchen re-furbished it to *Vega* 5B standard; installed a more powerful engine; added a large fuel tank; three different compasses; a drift-indicator and with a new registration NR7952, made it ready for an Atlantic attempt. NR-registrations were reserved for 'restricted' category aircraft such as those used for racing. Working at Teterborough airport, New Jersey, Balchen supervised Amelia's check of the aeroplane and taught her how to fly on instruments. Chasing the improving weather, Balchen flew the *Vega*, with Amelia and mechanic Eddie Gorski as passengers, via an overnight stop in St John, New Brunswick, up to Harbour Grace, Newfoundland, the departure-point for her flight. Following in Lindburgh's footsteps, the intended destination was Paris, France.

She took off at 7.15 pm (local) on 20 May 1932, the fifth anniversary of Charles Lindburgh's own trans-Atlantic feat. Encountering storms and icing conditions; strong winds from the north and miscalculating her drift; 2,026 statute miles and 14 hours and 54 minutes after take-off, on 21 May Amelia reached the coast of what turned out to be Northern Ireland. She had managed to keep the aircraft going despite a fuel-line leak; faulty altimeter and losing an exhaust pipe off the engine, so she abandoned going on to Paris, instead making a good landing in a pasture at Culmore, near Londonderry. Amelia Earhart had become the first woman – and only the second person – to fly solo across the Atlantic Ocean.

In 1930 George sold off his interest in the family publishing company and after a fire in November 1934, she and George moved out of Rye to set up home in Los Angeles, when George landed a job as chairman of the editorial board of Paramount Productions.

Amelia's reputation as a feminist icon – but she was never anti-men – was recognised in academic circles when, in late-1935, she was appointed by Purdue University College of Technology, in West Lafayette, Indiana, its consultant for women's careers and became a visiting faculty member and technical advisor to its aeronautical research department. Furthermore, in July 1936, two members of the Purdue Research Foundation donated a total of $40,000 ($\approx$ $905,000 in 2024) to allow Amelia to purchase a Lockheed 10-E *Electra*, twin-engine aeroplane, registration NR10620, to be used in the manner of a flying laboratory; enabling new aeroplane instrumentation to be tested under flying conditions. It was equipped with an auto-pilot; the latest compasses; drift calculators; radios; radio direction-finder; astro-dome; additional fuel tanks; up-rated engines and engine-performance monitors. This was a significant development for Amelia, who was quietly planning to undertake the ultimate challenge for any aviator – a round-the-world (RTW) flight and the Purdue Electra, with its array of up-to-date equipment, was an ideal machine for the purpose.

All this, however, had been leading up to a round-the-world (RTW) attempt by Amelia, ably assisted in the complex planning and logistics by husband George. The flight plan was to go round, from east to west, by the equatorial route. For the Pacific leg, tiny Howland Island (a coral dot with an area of just 1 square mile) 1,700 nautical miles (1,900 statute miles) south-west of Hawaii, with its recently established weather station and small landing strip, was to become a critical refuelling stop. The flight began in Oakland, California on 17 March 1937. On board the Purdue Electra was Amelia; her technical advisor and co-pilot Albert Paul Mantz (an air-racer and movie stunt-pilot), with Captain Harry Manning and former Pan American employee, Frederick Joseph 'Fred' Noonan, as navigators. Manning was a master mariner who met Amelia in 1928 aboard the United States Lines' SS *President Harding,* of which he was the captain. In addition to his navigational expertise, he was also an aeroplane pilot and was given a six-month leave to assist with this project. The flight plan was for Mantz to go as far as Hawaii; Noonan would continue as far as Howland Island and Manning would carry on to Darwin, Australia. Amelia intended to finish the rest of the flight alone.

The first leg from Oakland to Luke Field, Hawaii went smoothly and at 15 hours and 47 minutes, set a new record. Departure for the next leg, the long haul to tiny Howland Island, was delayed by bad weather. The weather improved and with a strong crosswind, Amelia lined up on the runway, opened the throttles and off they trundled. Gaining speed, the Electra swung slightly to the right. Amelia tried to correct this with the throttles rather than using the rudder. The aircraft swung to the left into a ground-loop. The undercarriage collapsed, dropping the aeroplane onto its belly in a shower of sparks. Loaded with 900 gallons of fuel, fortunately there was no fire. All four occupants emerged unhurt and it put paid to the RTW attempt, but Amelia was determined to have another go. The Electra was shipped to the USA where repairs were completed by mid-May.

Amelia re-planned the flight and now decided to fly from west to east. Paul Mantz was no longer interested and Captain Manning's leave had finished. Fred Noonan, however, agreed to navigate for the whole flight. Noonan was regarded as one of the most experienced marine- and air-navigators in the USA. After twenty years at sea, he made a new career in aviation, becoming highly respected as chief navigator for Pan American Airways, particularly during the development of its *China Clipper* routes across the Pacific Ocean. Sometimes Fred has been criticised for flaws in his character – but never for his navigation.

With George Putnam and a mechanic going along with them just for this first leg, Amelia and Noonan set off – without fanfare – from Oakland, California on 20 May 1937, heading for Miami via Burbank, Tucson and New Orleans; a distance of 2,332 nautical miles (nm). It was only when they reached Miami on 23 May, having proved the Electra was in good working order, that Amelia announced to the world that she and Noonan were already *en route* for their RTW attempt. The pair left Miami on 1 June bound for San Juan, Puerto Rico; Caripito, Venezuela; Paramaribo, Suriname; Fortaleza, Brazil and Natal, Brazil, which was reached on 4 June, after flying a distance of 3,387nm. The next leg crossed the Atlantic Ocean – at 1,727nm, the longest single distance so far – to Saint-Louis, Senegal, which was reached on 7 May. Between 8th and 14th June, Earhart and Noonan flew to Dakar, Senegal; Gao, French Sudan (Mali); Fort-Lamy, French Equatorial Africa (Chad); El Fasher, Anglo-Egyptian Sudan (N. Darfur); Khartoum, Anglo-Egyptian Sudan (Sudan); Massawa, Italian East Africa (Eritrea) and Assab, Italian East Africa (Eritrea), a total of 3,714nm, across Africa to the mouth of the Red Sea. Everything was going well so far, although Noonan found navigation tricky because his charts were often inaccurate.

FATAL FLIGHTS OF THE RICH AND FAMOUS

The leg from Assab to Karachi, India (Pakistan), on 15 June, was made in one hop of 1,627nm. Calcutta (Kolkata), on the east coast of India was a further 1,178nm, and that was reached on 17 June. At each stop, Amelia cabled her view of events and progress to George, who would then communicate this valuable story to the world's media.

Over the next three days they flew 2,195nm, to Akyab (Sittwe), Burma (Myanmar); Rangoon, Burma (Myanmar); Bangkok, Siam (Thailand); Singapore, Straits Settlements (Republic of Singapore); arriving at Bandung, Dutch East Indies (Indonesia) on 21 June. Here, they were delayed by monsoon weather and it was 24 June before they could continue to Surabaya, Dutch East Indies (Indonesia). Encountering some problems with the fuel mixture and engine control equipment, they had to retrace their steps to Bandung on 25th since it had the only workshop able to repair these instruments. Onward, once more they flew to Surabaya on 26th then to Koepang on Timor Island (Indonesia) on 27th and Port Darwin, Australia on 28th. They left Darwin on 29 June, bound for Lae, 1,000nm ahead in New Guinea, which was reached safely.

Friday, 2 July 1937; 10.00 am, local time and already the air was heating up. This was to be the longest and most testing leg of the whole journey. The special fuel tanks were filled to capacity with 1,100 gallons of 87-octane; plus 100 gallons of 100-octane petrol in the standard wing tanks – vital in order to lift the heavily over-loaded aeroplane off Lae's dirt runway. All un-necessary kit was removed and packed for shipment home. All vital equipment was checked and serviced. News stories were cabled. Two US Navy ships: *Ontario*; *Myrtlebank* and the US Coast Guard cutter *Itasca*, were in position *en route* to assist with navigation by radio. Tiny Howland Island, 2,227nm, and 18 hours away beckoned and after 19,000nm (22,000 statute miles) completed so far to Lae, there was only another 7,500nm (8,600 statute miles) to California and home.

Amelia Earhart and Fred Noonan never reached Howland Island. They just disappeared without trace, possibly within 100 miles of the island, but no-one really knows since, in the end, radio communication in both directions failed or was too inadequate to be of help at the critical point of the flight. Over the years, despite fruitless searches and many conspiracy theories concerning their fate, 'mythistory' began to take over. As Lutz asserts:[3]

> 'Uncertainty itself begets rumours. Unsatisfying "answers" and unresolved mysteries mark the birth of speculation and conjecture. If Earhart's disappearance continues to remain unsolved, theories (and myths) will exist that attempt to "explain" the unexplained. This phenomenon is consistent with other rumours that emerge after a "great" man or woman's unsatisfying, unexplained, or sudden death.'

The official US government line is the 'Crash and Sink Theory': that they ran out of fuel and were lost when the aeroplane crashed into a shark-infested sea. More recently, an alternative, private, version was put forward: The 'Nikumaroro Island Theory.' Never proven, it was at least a more reasoned approach than the most enduring conspiracy theories that centre around the 'Japanese Capture Theory' – which itself developed two strands: the 'Saipan Theory' and the 'Irene Bolam Theory.' It is not intended to delve into these here, since they are both convoluted and freely available all over the Internet – and detract from the real, verifiable, achievements of this aviation legend.

The latest development came as recently as January 2024, when a private marine exploration group claimed to have sonar-imaging evidence of what is alleged to be the wreck of an aeroplane resembling an *Electra*, on the seabed at a depth of 16,000 feet – deeper than the *Titanic* – about 100 miles west of Howland Island.

Was this intriguing mystery about to be solved at last?

Well, sadly the answer that was announced to the world's press in December 2024 was 'No.'

When a submersible camera was lowered to this site, the images it sent back showed the 'wreck' to be just a rock formation that looked deceptively like an aeroplane.

So, the hunt continues.

Chapter 5

Amy Johnson

5 January 1941

For women in the UK, achieving prominence in aviation during the inter-war period was generally the preserve of those from wealthy or titled backgrounds. Amy Johnson was from neither of these but, through sheer determination she rose above all others to become one of the most inspirational of the twentieth century's female aviators.

Born in Hull on 1 July 1903, in her early life there was no indication that she had any interest in aviation or in becoming a pilot herself. If anything, there ought to have been a connection to the sea, since her father, John William Johnson (known as 'Will'), worked in the family fish sales business.

Initially, Amy and her sister Irene were educated at a small private school on Anlaby Road, near the family home but, around the age of twelve she went to a local municipal secondary school. Passing through a 'tomboy' stage, in 1917 she not only lost an argument with a cricket ball, she lost her front teeth to go with it. The problem was fixed with dentures but, thereafter, Amy always remained self-conscious about them. At school she was good at French and aged 17, Amy passed her School Certificate and Cambridge Senior local examinations and was encouraged by her parents to remain in the sixth form until she was 19 years of age. In 1922, with the promise of a local education grant to cover university fees for students who committed to a career in teaching, she applied successfully for a place at Sheffield University to study modern languages.

Just prior to leaving for university, Amy was introduced by an aunt to Hans Arregger, a young Swiss businessman working in Hull. Immediately attracted to his good looks and charm, Amy became besotted with him, spending all her spare time in his company. Hans was eight years older than Amy and while content to be her boyfriend – and eventually her lover – he always avoided making any other commitment to their partnership. There is archived evidence of Amy's 286 letters to Hans during their six-year relationship; the result of Hans keeping all her letters for the rest of his life – which, in view of his apathy about the future of their relationship, suggests he must have cared something for her nonetheless.

In 1925, possession of an ordinary BA degree in French, Latin and economics did not open doors to employment as easily as Amy imagined it would. Her dreams of travel to foreign lands, high wages and married bliss were just that: dreams. Having attended classes in shorthand, typing and general secretarial duties, Amy picked up her first job as a shorthand-typist with an accountant in Hull.

Having moved to London in April 1927 as a sales-trainee in the John Lewis/ Peter Jones department store in Sloane Square, in her longest letter to him, dated 25 March 1928, Amy severed her deep commitment to Hans. By mid-summer 1928, it was all over between them; Amy had quietly decided she was going to find a new avenue for her dreams – and it should be one she could control alone. In an era when aeroplanes, aviation, pilots and adventure constantly filled the public's eye and the world's media, Amy Johnson decided she wanted to learn to fly. On 28 April 1928, she hopped on a bus to Hendon then on to Stag Lane, Edgeware, where she had heard that there were aeroplanes to fly.

Stag Lane was home to the London Aeroplane Club and flying school and Amy's new life began on 15 September 1928, when she had her first half-hour flying lesson with Mr F.R. Matthews, a junior instructor, in a De Havilland (DH) 60 Moth. She also had instruction from the chief flying instructor (CFI), Captain Valentine Baker MC AFC an ex-Royal Flying Corps (RFC) fighter-pilot veteran of the First World War. Amy flew mainly with Captain Baker, who made her practise landings time after time, with the aim of reaching the holy grail of a neat three-point landing – and no kangaroo-jumps. Landings, though, would never be Amy's strong point!

Her dual-time crept up to 15 hours and 45 minutes before finally, on 9 June 1929, she went solo. It was clear that Amy was not what might be regarded as a gifted or skilled pilot but her competency, coupled with personal qualities of determination, will power and a great desire to see things through, made her the pilot the world eventually saw. Furthermore, Amy took a keen interest in the technicalities of aerodynamics and engines. Her enthusiasm caught the attention of the chief ground engineer, Charles H.G.S. 'Jack' Humphrey. They became good friends, with Jack actively encouraging her interest in engine and airframe maintenance. It is suggested that, while chatting to Jack Humphrey and Valentine Baker, Amy may have had the notion of flying to Australia planted in her mind.

Amy's life now focussed almost entirely on aeroplanes and her ordinary job began to fade in its attraction – apart from the wage it brought in of course! Determined to study for an aero-engineering qualification, she persuaded the London Aero Club secretary and Jack Humphrey to take her on as an unofficial, unpaid, engineering apprentice. In addition to learning how to strip engines; maintain them and put them back together again (with no bits left over!), Amy did more solo flying now and with 19 hours in her log-book, took her oral exam and flying tests in a club DH60X Cirrus Moth on 28 June 1929. Her RAeC aviator's certificate, No. 8662, allowed her to apply for an Air Ministry private pilot 'A' licence, No. 1979, issued on 6 July, permitting her to fly all types of flying machine – but not for remuneration.

It is a measure of her grim determination that on 9 December 1929 Amy was awarded her 'C' ground engineer's licence, the first woman to receive one issued in the UK, authorising her to carry out pre-flight inspections of aero engines, such as *Cirrus*, *Hermes* and *Gipsy*, the types usually found on the De Havilland Moth and similar light aeroplanes. In March 1930 she qualified for her second-class navigator licence, an essential preliminary to the award of the 'B' pilot licence – for which she still had to accumulate many flying hours on solo cross-country and night flights to reach the required 100 hours.

The Amy Johnson we are seeing here is a far cry from the earlier, somewhat vacuous, moody younger version and yet, now she had found a purpose in life, she seemed to thrive under pressure and her health – so far – seemed to be holding up under the strain.

The idea of making a landmark flight to Australia – in the interests of both herself; women and the future of aviation – was not the result of a single 'lightbulb moment,' but rather it blossomed in Amy's mind after that first casual conversation with Baker and Humphrey. In a way, it gave a focus and purpose to all the hard work she was putting in for her air and ground licences and fitted into her new-found mindset about a woman capturing a significant air record – a *Blue Riband* of records: London to Australia; solo. So far, Bert Hinkler AFC DSM, was the only person to make the 11,000-mile journey, solo, in an Avro Avian land-plane in 1928 and she set her sights on his 15½ day record, too.

Amy hoped to begin this immense undertaking in early May 1930. How to pay for it all, though; that was the big issue. She needed at least £1,000 (≈ £54,000 in 2024) for an aeroplane and other costs. Her father offered £500 and with some intricate 'networking' the *Castrol* oil magnate, Lord Wakefield agreed to find the rest. She bought a two-year-old DH Moth powered by a 100hp, Gipsy engine, registration: G-AAAH, for £600, had it painted green and named it *Jason*. It was fitted with a heavy-duty propeller – which lowered the cruise speed to about 85mph – and she had a spare prop lashed to the outside of the fuselage with tools, spare fabric and dope inside. When organising the intermediate refuelling locations, the apocryphal story goes that Amy simply laid a ruler on a map between London and Darwin and said that was her route. The Wakefield Oil Company then sorted out all the *en route* airfields and local arrangements for fuel, maintenance, airspace- and landing-permits.

Dressed in jodhpurs, shirt and jumper under a warm *Sidcot* overall-suit and wearing a parachute, Amy, watched by a small group that included her father, Jack Humphrey, Winifred Irving and some Wakefield staff, set off without fuss, from Croydon airport at 7.30 am on Monday, 5 May 1930. This novice pilot, with just 99 hours solo in her logbook, was on her way to Australia!

On Saturday, 24 May 1930, Amy Johnson, aged 27 years, touched down at Fannie Bay airfield in Darwin, Australia, 19½ days out from England. She had not beaten

Bert Hinkler's record, but that pales into insignificance considering she had only missed out by four days and – more to the point – had completed the finest solo flight ever made by a woman pilot. It is impossible to over-state the extent of her acclaim by the British nation, its public, her parents, Jack Humphrey and – not least – by the rest of the world. Amy was made a Commander of the Order of the British Empire (CBE) just ten days later and received many other special awards for her technical achievements including the Harmon Trophy for women aviatrix of 1930. The *Daily Mail* continued its support of British aviation by paying Amy £10,000 (≈ £545,000 in 2024), in addition to £2,000 (≈ £109,000 in 2024) for her serialised story.

After just a day's rest, Amy embarked on a gruelling six-week public tour of Australia before returning to England more leisurely by ship to Port Said then by airliner to England. *En route*, Amy stayed several days in Aden where on 24 July she visited No. 8 Squadron at RAF Khormaksar.

Upon arrival in England, she faced a vast public welcome, with an estimated one million people lining the twelve-mile route from Croydon airport to a reception at Grosvenor House Hotel in central London. This was followed by a gruelling UK air-tour programme. It put an immense strain on her physical and mental health and on medical advice, she took a holiday and later had to take another rest in a nursing home, followed by dropping out of the limelight to stay quietly with friends. Her faithful DH 60G Moth, G-AAAH, *Jason* was saved for the nation and can be seen in the Science Museum, London.

The pressure of fame began to tell on her again. That winter, during another hectic lecture tour, Amy's health broke down and she went into hospital, followed by a long cruise to Madeira and South Africa and back, to recuperate. It was here that the flight-paths of Amy Johnson and Jim Mollison first crossed in March 1932.

Amy and Jim met again in London and on 9 May 1932, Jim popped the question and Amy said 'yes,' heralding a period of living the 'high life' based at the Grosvenor House Hotel. Although the value of being celebrities brought them 'freebies,' 'perks' and fees, their celebrity lifestyle was a constant drain on their income and prize-money. They married in a quiet ceremony on 29 July and now, as the most news-worthy duo in the land, their married life was lived in the full glare of publicity. Amy learned to dress elegantly in the latest style and jewellery – provided, sometimes freely, by fashion houses such as Schiaparelli and Chanel – and also learned that public exposure translated into income.

For something different Amy and Jim planned a tilt at the long-distance, non-stop record, using a new twin-engine, biplane airliner; a DH 84 Dragon, G-ACCV, named *Seafarer*. They would make a flight from Croydon, London to New York departing on 8 June, followed by the distance attempt: New York to Baghdad. The project failed at the outset, when the overloaded Dragon hit a gully and its undercarriage collapsed. Undaunted, the Dragon was repaired, but the departure point was moved to the long, hard beach at Pendine Sands, Wales. This time, all went well during take-off at noon on 22 July. Sighting the Newfoundland coast, New York, their objective, was still 1,100 miles away. Headwinds gobbled-up their fuel and with New York possibly un-attainable, after thirty-nine hours in the air they crash-landed at Bridgeport, Connecticut on 24 July 1933. Both were hospitalised but upon recovery, were the subject of immense attention by the New York social scene – particularly Amy, who sparkled in the limelight. Jim, the eternal playboy, like to be 'top dog' and made no secret of his jealousy – he, too, was feeling the pressure of constant publicity and the marriage was showing the strain. America perhaps put the final seal on the break-up of this marriage of two very independent and strong-willed people.

While convalescing, Amy and Jim were royally entertained by George and Amelia (Earhart) Putnam at their palatial home in Rye, NY, during which time they also had a ticker-tape welcome in New York and met President Roosevelt. Jim returned to England in August, while Amy stayed on with the Putnams, thoroughly enjoying herself in Amelia Earhart's company.

Jim proposed that they both enter the MacRobertson Air Race from England to Australia. The race commemorated the centenary of Melbourne and the state of Victoria and an Australian businessman, Sir MacPherson Robertson, put up a £15,000 (≈ £895,000 in 2024) prize, which drew huge international interest from the aviation world. For this competition, due to begin on 20 October 1934, from Mildenhall, Suffolk to Melbourne, Victoria, the Mollisons would use a new DH 88 Comet, G-ACSP, a twin-engine, two-seat racer, named *Black Magic*.

When Jim went off to America in 1935, Amy sought the sun and rest in Madeira. Their lives were now incompatible and Jim's alleged womanising finally brought about virtual separation by the end of the year. Amy decamped to Paris and started divorce proceedings in September 1936. By the close of the year, her life alternated between London, New York and Paris; clothed now by Main Rousseau Bocher (Mainbocher), couturier to nobility and the stars; she crash-landed her Beech B17L Staggerwing, G-ADDH, aeroplane while commuting to France (Amy was UK agent for the Beechcraft Company); more ventures and ideas fell by the wayside; she took holidays in the south of France and finally had her teeth fixed in New York. This hectic life-style took its toll and in January 1937, Amy's doctor packed her off for rest at the idyllic Clinique Valmont in Glion-sur-Montreux in Switzerland.

Amy's divorce was made absolute on 24 August 1938 and as a single person, enjoying a very comfortable life-style – but no 'proper' job – Amy's finances got into a mess again. She was mortified when, after approaching the Air Ministry for employment, she was offered a clerical post at £5 a week! However, in June 1939 she landed a job with Portsmouth, Southsea and Isle of Wight Aviation Ltd., as a professional pilot (at last!). Her work included ferrying passengers to and from the Isle of Wight as well as flying by day or night as a passive target for Army anti-aircraft searchlight installations in the Portsmouth and Southampton district – a sign of the imminent war. When war was declared, the company was taken over by the Air Ministry and its operations – and Amy – moved to Cardiff airfield for Army co-op work and communications duties, which even included her flying over to RAF and civil airfields in France during the 'Phoney War.' When this contract fell through, in May 1940 Amy joined the new women's section of the Air Transport Auxiliary (ATA) and at the age of 37, with 2,300 flying hours to her name, Amy donned the uniform of an ATA ferry pilot – and was never happier.

Based at No. 1 Ferry Pool at the White Waltham airfield ATA HQ, most of her flying was delivering or collecting pilots after they had moved RAF aircraft – mostly trainers at this time – around the country. This is where the taxi-planes came in and Amy took her turn to act as taxi-pilot, usually in an Avro Anson, as well as delivering individual aeroplanes herself.

It was on an individual ferry flight, taking Airspeed Oxford II, V3540, from Prestwick, Scotland down to No. 15 Service Flying Training School (15 SFTS) at RAF Kidlington, Oxford, that Amy Johnson (in 1939 she reverted by deed poll to her maiden name) went missing on Sunday, 5 January 1941. Initially, Amy was to fly an Airspeed Oxford from the Hatfield factory airstrip to Prestwick on Friday, 3 January. Due to bad weather, she landed at RAF Tern Hill, Shropshire and stayed overnight in a nearby hotel. Next day, Saturday 4th., Amy was able to reach Prestwick to deliver that aircraft safely. She was then tasked with picking up a different Oxford (V3540) at Prestwick and flying it down to Kidlington airfield, near the city of Oxford. Leaving Prestwick at 4.00 pm that same afternoon, in cold weather and an overcast sky, Amy made a precautionary landing at Squires Gate airfield in Blackpool, staying Saturday night with her sister Molly and her husband who lived in the town. In her book, Constance Babington Smith records Amy mentioning to the couple: 'that she thought the compass of her aeroplane was a little out of adjustment' but declined to report it, since she wanted to get on with her journey without further delay. Leaving her sister at 10.00 am on Sunday, low cloud still persisted and some mist rolled in as she waited in the aircraft to see if it would improve. It did not and she departed Blackpool at 11.50 am, planning to climb above the cloud for the one-hour flight to Kidlington. By 5.00 pm, with no word from Amy, back at White Waltham they were getting very worried. What happened next shows she was certainly off-course. It suggests her being unable to find a gap in the cloud cover and baling out having judged she had flown as far as possible without finding a gap, but before reaching the sea.

Giving evidence at the inquest on his captain, Lieutenant Henry Patrick O'Dea Royal Naval Reserve (RNR), recalling events on board the mobile barrage-balloon vessel HMS *Haslemere*, said his ship escorted convoy CE21 from St Helen's Roads, Isle of Wight on 4 January and was due to arrive at Southend on 5 January. The weather was bad, with snow showers. It was extremely cold and there was a big swell on the sea. Having just rounded North Foreland, he was on the bridge of his ship off Herne Bay on Sunday 5 January when, at 3.30 pm, about a mile ahead he saw a parachute descend and an aeroplane circling downwards towards the water. His ship and another escort, altered course to go to the aid of the parachutist who was now in the water. The aeroplane had also glided down onto the surface of the sea. Closing to within half a mile of the person in the water, HMS *Haslemere* ran onto a sandbar. In the ten minutes that the ship went astern and pulled clear, the person drifted quickly towards it on the fast current and from the shouts for help, he 'knew that person was a woman.' Several seamen threw lines but all missed or were not grabbed. Able Seaman Raymond Dean said he climbed over the ship's side and crouching on a ledge above the stern, stretched out a hand but could not reach. He said he heard the person crying for help and it sounded like a woman's voice. 'She was wearing a flying helmet,' he said, 'and she drifted under the stern.' Seaman Nicholas Roberts said he, too, thought it was a woman and that 'she' drifted under the stern of the ship when it lifted on the swell; when it dropped, no-one saw that person again.

All three witnesses claimed they saw a second 'person' in the water some distance away from the 'woman.' This was when *Haslemere's* captain, Lieutenant Commander Walter Edmund Fletcher RN, left the bridge, threw off his overcoat and dived into the sea. He swam to the farthest 'person' and seemed to support it briefly before letting go and trying to swim to a ship's boat. Did he find it was not a body after all? In the freezing conditions, he was pulled unconscious from the water and died later in hospital having never regained consciousness. For his bravery that day Lieutenant Commander Fletcher was posthumously awarded the Albert Medal.

A valise bearing her initials 'A J' (some reports suggest two bags were found) and containing Amy's possessions including, clothing; her log-book and other personal documents, were picked up from the sea near the aeroplane wreckage by sailors from another convoy escort, HMS *Berkeley*. The bag (or bags) most likely floated out of the fuselage door, opened when Amy baled out. The mystery second person – a body for which was also never found – spawned controversy and conspiracy theories, none of which have been definitively resolved to this day. Was there a second parachute? Was another, unauthorised, person in the aeroplane? Was the 'body' one of the floating bags? Whatever the multitude of theories for a 'second person,' from eye-witness evidence it seems clear that Amy floated very close to the ship and it seems probable that she died after being struck by the hull of HMS *Haslemere* or by the ship's propeller and was either killed by the blow or it caused her to be drowned. The body of First Officer Amy Johnson CBE (37), Air Transport Auxiliary, was never found and she is commemorated on the Runnymede Air Forces Memorial to the Missing; while statues of Amy are to be found in Herne Bay, the official location of her disappearance, and in her home town of Hull.

Chapter 6

Carole Lombard

16 January 1942

Heads or tails? The simple toss of a coin decided if Carole Lombard would travel home to Hollywood by air or by rail. Her husband's personal MGM publicist and great friend, Otto Winkler, wanted to go by rail; her mother, Elizabeth, detested flying but Carole – who viewed aeroplanes a bit like taking a cab – was anxious to return to her husband, Clark Gable, as quickly as possible and thus favoured flying home. The story goes that to settle their friendly argument, they agreed to flip a coin – and Carole won. She won the toss, but lost her life. The aeroplane in which the three of them made the subsequent trip, crashed into a mountainside, killing all twenty-two souls on board. At the time of her death in 1942, vivacious blonde movie-star Carole Lombard was at the pinnacle of fame, both in her own right and as one half of the most- famous celebrity couple in Hollywood: Lombard & Gable – the 'Jolie & Pitt' of their day.

Born plain Jane Alice Peters, on 6 October 1908 into a wealthy family in Fort Wayne, Indiana, her parents' marriage was not a happy one. In 1915, Jane's mother, Elizabeth, separated from her father, Frederick and moved with Jane to the Wilshire district of Los Angeles.

Despite a few disappointments along the way, when she reached the age of sixteen, her ash-blonde hair; sparkling blue eyes and sexy visual potential was spotted by a producer from Fox Films and after a screen test, Jane was signed up by the studio at \$65 a week (≈ \$1,000 or £750 in 2024). Blossoming into a very attractive and determined young woman, when her later, quirky comedic talent shone through, it propelled her to on-screen success all over the world. Pressed by the studio to change her rather ordinary-sounding names to something more enigmatic, she chose the combination of Carol (after random mis-spellings in film billing and by the press, she decided she liked the extra 'e' and officially added it in 1936) and Lombard, both names garnered from her circle of friends.

Her first brush with the Grim Reaper came in 1926 when, as a passenger in a car driven by a 16-years-old boyfriend, Harry Cooper, they were in a bad accident after a night out on the town. Quite apart from suffering painful injuries, Carole's face was ripped by shards of glass from the shattered windscreen. Death passed her by this time, but his price was a horrific gash around her left eye and cheek that needed twenty-five stitches.

Her first Hollywood sound film was *Wanted* (Pathé; 1929), with *High Voltage*; *Big News* and *The Racketeer* quickly following for this studio during the year. In 1930, Carole left to try her luck in New York and netted a contract with Paramount Pictures and while at Paramount, its top male actor, William Powell, stepped into her life. She had made eight films for the studio by the time they co-starred in *Man of the World* in 1931 and the pair hit it off so well that they decided to get married; he was aged 39 and she: 22 – it would be fun; but doomed to failure! For all her angelic charm, it seems incongruous to record, that this lovely actress had acquired a solid reputation as a cigarette-smoking, foul-mouthed woman who actively employed the 'f'-word and other juicy expletives in public. Nevertheless, she was without doubt a naturally kind and considerate person who loved the party life and viewed the prospect of a life of domesticity with no enthusiasm whatsoever.

Powell and Lombard made another film together: the romantic drama *Ladies Man* (Paramount; 1931) then, accepting they were incompatible, divorced amicably in 1933. They co-starred in one movie after they were divorced: *My Man Godfrey* (Universal; 1936). It was a huge box-office success and although nominated for six Oscars, including one for Carole as Best Actress, sadly the film won none – though Carole was runner-up in her category. It did, however, firmly establish her as a first-class film comedienne in the Hollywood film genre known as 'screwball comedy.'

Carole was now a sure-fire box-office draw and saw her salary climb from $300 a week (≈ $5,400 or £4,300 in 2024) when she signed her first contract with Paramount back in 1930, to $3,500 a week (≈ $80,000 or £64,200 in 2024) by the mid-1930s. Turning out at least two or three movies a year her talent brought wealth, jewels, fashionable clothes and a grand life-style, but marital bliss still eluded her. After William Powell, she had a brief romance with a handsome singer-songwriter-violinist-actor named Russ Colombo (or Ruggiero Eugenio di Rodolfo Colombo, to give him his full name).

It is a measure of Carole Lombard's high status throughout 1930s Hollywood that she was cast as leading lady with many of the enduring male icons of the silver screen, such as: Gary Cooper, Clark Gable, Charles Laughton, Fred MacMurray, Frederick March, George Raft and James Stewart. In 1936, Carole Lombard and Clark Gable began what became one of Hollywood's most famous real-world love stories. Having had little or no professional contact since a movie in 1932, they met at a party, hit it off and began dating; Carole was still single and Clark was separated from his wife, Ria Langham. The romance blossomed; Clark divorced Ria and Lombard and Gable were married in a very low-key wedding at the end of March 1939.

Following the attack on Pearl Harbour, on 8 December 1941 the USA declared war on Japan and on 11 December, entered the war against Germany on the Allied side. This was a cue for many patriotic actions by celebrities, ranging from enlisting in the armed forces to national fund-raising. Carole's husband, Clark Gable, already a prominent member of the Screen Actors Guild, was a natural choice for chairman of the Hollywood Victory Committee (HVC), an organisation set up just a few days after Pearl Harbour, to co-ordinate the efforts of film, stage and radio celebrities to raise funds and entertain the armed forces for the benefit of the nation. One of the first requests made by the US government to the HVC was to send a film star to tour Indiana for three days, promoting the state's campaign to sell US Defense Bonds (later called War Bonds) to the public. The request came to Gable via Howard Dietz, head of Public Relations for Paramount. Gable claimed he was unable to oblige personally but who else could he nominate but his wife Carole, who was actually born in the state of Indiana – a real live 'Hoosier'! Enthusiastically, she agreed to take on the task, believing it to be nothing less than her patriotic duty.

On Monday, 12 January 1942, Carole made the long, three-day journey by train from Los Angeles to Indianapolis, by way of Salt Lake City and Chicago. She was accompanied by her mother, Elizabeth and by Otto Winkler, Clark Gable's personal MGM publicist and a staunch friend of the family. Their train stopped briefly in Salt Lake City, where she made a speech to a crowd assembled at the station, exhorting them to do the patriotic thing and go out and buy War Bonds. On Wednesday, 14 January: Chicago – yet another speech; meetings with US Treasury officials about the mechanics of selling war bonds and a visit to the *Chicago Tribune* offices. Carole was interviewed on the *Tribune's* own radio station: *WGN* and she had formal colour portraits taken in the newspaper's photographic studio. Meanwhile, Elizabeth had gone ahead to their hotel, where she met family relatives, in from Fort Wayne.

Carole had a very heavy schedule on Thursday and was determined to cope with it by getting her beauty sleep in. Against all official advice and over-riding Otto's protestations, Carole insisted on both of them flying down to Indianapolis that evening, after she had completed all her engagements. Elizabeth would follow on by train, with their baggage, the next day, Thursday, 15th.

At Indianapolis Union station around 1.30 pm on the 15th, they met Elizabeth off the train and Carole, looking every inch the Hollywood super-star, was formally greeted to the city by the Governor, the Mayor and other dignitaries. Now the rest of the day was going to be a fast-paced, gruelling programme at various venues across the city, making speeches and selling bonds hour after hour. Her first engagement saw Carole raising the national flag, making a speech and selling bonds at the Indiana Statehouse. Then, on to the Claypool Hotel to raise another flag and open an armed-forces recruiting centre in the foyer – and sell bonds. Next came a reception at the governor's mansion; a high-tea and social event to mingle with, thank and encourage the women of the Indiana Defense Savings Staff. Now it was back to the Claypool for a formal evening dinner mixing with the great and good of Indiana, before leaving for the climax of the day: a rally at the Cadle Tabernacle. Many were the costume-changes that day, but all the time Carole looked every inch the beautiful Hollywood star. The crowds, running into thousands, adored her – and she had found her hotel room absolutely filled with red roses from Gable; a gesture that re-energised her. Everywhere she went, at bond sale events all over the city, the money came rolling in, greatly exceeding expectations. By the end of her days work, she had helped to raise a staggering $2,000,000 (≈ $39,700,000 or £31,700,000 in 2024). With the Second World War for the USA just thirty-nine days old, Carole Lombard was the first big Hollywood celebrity to undertake a fund-raising bond tour in this new conflict and it was a very high standard that she set for future events by her fellow stars.

Around 8.30 pm, in that long and tiring day, still – apparently – effortlessly looking at her most glamorous, came the climax of the tour. A huge patriotic rally was held in the cavernous Cadle Tabernacle; yet another speech, bands playing, soldiers marching and then rounding off the event by leading the packed house – estimated at 12,000 people – in an emotional rendition of the *Star-Spangled Banner*. She held the audience in the palm of her hand! This iconic moment, with Carole dressed in a gorgeous strapless evening gown; arms outstretched and looking like a million dollars, was captured forever in many iconic photographs.

It was gone midnight when the weary trio, Carole, Elizabeth and Otto, packed their bags to prepare to head back to Los Angeles on the 16th. Otto and Elizabeth were all for staying overnight and then catching a train back to California later in the new day. Neither liked the idea of flying and indeed Otto had been briefed by Gable; people in the film industry and the federal government, to avoid the risk of flying at all costs. Carole though, who had given her all that day and was exhausted, was not in the mood to face another three-day journey by train, interspersed with whistle-stop speeches *en route*. Despite all his pleas and protestations, she instructed Otto to go and buy three air tickets for the first available flight – and no arguments!

Why she was so adamant about flying home has been the subject of much speculation that has never been definitively resolved and of course, those who could do so are no longer with us. Clark and Carole are said, for example, to have had a disagreement prior to her departure on the trip and she was keen to return 'to mend fences' with her husband. Much is made of Clark's dalliances with his co-stars and how Carole, aware of the rumours, was anxious to get back to him. But Carole was not beyond flirting with her young co-stars either.

One of the most enduring myths, though, is that it was the toss of a coin between Carole and Otto that settled the argument, over train or plane, between the three of them – it is claimed Otto lost the toss and against his better judgement, did as Carole demanded. Due to some cancellations, he managed to buy three tickets on Transcontinental & Western Air (TWA) Flight 3, due to arrive in Indianapolis around 4.00 am Central Time (CT) on Friday, 16 January, 1942. When appearing in print, the provenance for the coin-toss story is always incomplete. The incident is said to have taken place between Carole, Otto and Elizabeth when the three of them were alone in their suite, after all the guests had departed. So, unless Otto confessed his weakness to someone, how did this story enter the public domain? No-one has ever answered that question.

Shortly after 1.00 am on the 16th, Carole, Otto and Elizabeth were driven to the municipal airport to catch their flight. TWA '*Sky-Apache*, Flight 3' originated at La Guardia airport, New York and was a regular, daily service, using a Douglas DC-3-382 '*Sky-Club*', twin-engine airliner, with seats for up to 21 passengers in its day-time configuration, in an un-pressurised cabin. This run was being operated with an aircraft registered: NC-1946 and bearing the airline number '387' on the fin. The schedule for the 2,635-mile route planned to reach Burbank, Los Angeles, CA, in seventeen hours – weather permitting. Take-off time in New York was 11.00 pm, Eastern Time. There were intermediate, optional, stops for fuel, passengers and freight at Newark, NJ; Pittsburgh, PA; Columbus, OH; Dayton, OH; Indianapolis, IN; St Louis, MO; Kansas City, KS; Wichita, KS; Amarillo, TX; Albuquerque, NM; Winslow, AZ; Boulder City AZ; finally arriving at Burbank, in the heart of Hollywood.

Shortly after 4.00 am Central Time (CT), the DC-3 landed at Indianapolis; loaded Carole's party; took on fuel and was airborne again at 4.27 am. Upon landing at St Louis, progress was delayed for two hours due to fog. Allowing for small amounts of time lost shuffling passengers and freight at each intermediate airport; another bad-weather delay at Kansas City; together with some headwinds, by the time Flight 3 reached Albuquerque at 4.06 pm Mountain Time (MT), it was running over three hours late. The original flight crew from New York was changed, first at Columbus, where Captain John Hagins took command and then Kansas City, where Captain Ernest Pretsch took over. Passengers, too, had come and gone as the aeroplane progressed down the route. Now another new flight crew took over at Albuquerque, comprising Captain Wayne C. Williams; First Officer Morgan A. Gillette and Air Hostess Alice Frances Getz.

Normally, the flight would depart Albuquerque and not land again until Burbank, but today would be different – and maybe this is where it all started to unravel. Before the seven civilian passengers disembarked prior to refuelling, the TWA agent went aboard to inform them that, a group of US Army Air Corps (USAAC) airmen claimed war-time priority seating for the trip to Burbank. These men, assigned to USAAC Ferry Command, had delivered Army aeroplanes from their Long Beach base to Montreal, Canada – where they were destined for England. The airmen were returning to their base and they and their kit – including parachutes – had priority over any civilians. All civilian passengers must therefore disembark and wait for the next available flight.

Carole was incensed by this. Probably letting rip with some salty words, she refused to accept the decision; using her status, not only as a national celebrity but also as someone who had just bust her gut on a Federally-sponsored war bond fund-raising tour. She insisted her party had justifiable top priority and as a result of her forcefulness, the agitated agent relented and Carole, Elizabeth and Otto were permitted to re-board. But, the other civilians – with the exception of just one other: Mrs Lois Hamilton, wife of a serving USAAC officer, flying out to join him – were bumped off the flight. The 'lucky' ones were: Joseph Szigati, a Hungarian violinist, reported to be going to Hollywood to work on a new film with Irving Berlin; Miss Mary Johnson of Benica CA; Mrs Florence Sawyer of Portland MO and Mrs Carl Brandoner of Holton KS. These four continued on a flight leaving the next morning.

With three airmen already aboard (from Amarillo), twelve more airmen were ushered into the cabin and it is likely the aircraft was now slightly overloaded. There is a suggestion of some 'creative' number-crunching by the hard-pressed TWA passenger agent at Albuquerque, to get the payload weight down to within authorised limits. Although two seats remained unoccupied on NC-1946, the calculated weight of the overall passenger and freight load – considered in conjunction with the resultant reduced fuel load – would not permit any more passengers to be carried. Baggage was secured on the empty seats and at front and rear of the cabin, as well as in the usual freight spaces. If there were no passenger/technical or crew reasons to land at an intermediate airfield, it was quite usual for this flight to fly non-stop from Albuquerque to Burbank. However, on this occasion the aircraft weight would only allow sufficient fuel for flight as far as Boulder City – plus the usual 'diversion-reserve.'

This mess took time to resolve, but Flight 3 left Albuquerque at 4.40 pm MT (3.40 pm Pacific Time, [PT]). After take-off from Albuquerque, with delay-time mounting, Captain Williams radioed his HQ at Burbank for permission (received) to cut out the stop at Winslow and press on to Boulder City. It was clear, though, that delays were pushing them towards darkness. There was no prospect of him landing in daylight at Boulder City – and it had no runway lights. Following company procedure therefore, Captain Williams, having also radioed Burbank for permission to go direct to McCarran Field, Las Vegas, NV – the approved TWA alternative airfield equipped with runway lighting – received approval and now replanned his route. McCarran Field was a USAAC station (Nellis Air Force Base in 2024), about 20 miles north of Boulder and just outside Vegas. It had an arrangement with TWA, permitting the company to use it for commercial flights and Williams radioed McCarran, advising them of his impending arrival. Three hours later (at 6.36 pm, PT) the aeroplane landed at McCarran Field. More fuel was taken on and TWA Flight 3, with the same passengers, departed McCarran at 7.07 pm (PT) – next stop Burbank; end of the line.

Fifteen minutes after take-off, climbing out at about 130 mph in darkness, NC-1946 flew into a vertical rockface 7,770 feet up on Potosi Mountain in the Spring Mountain range. The DC-3 was destroyed and no-one was found alive.

It was a nightmare for would-be rescuers to reach the extremely remote site of the crash, which was also above the snow-line and in near-zero temperatures. Fire-damaged debris and human remains were scattered down several hundred feet of a near-vertical slope below the impact point and it was several days before the passengers and crew could be recovered and identified (see Appendix 2).

The impact point was 33 miles on a bearing of about 215 degrees magnetic from the airport – which is roughly consistent with the headings written on the flight-plan. The flight-plan was to level off at 8,000 feet altitude; Mount Potosi was 8,500 feet high and the top ledge of this particular cliff face was at 7,850 feet. Flying such a heading and altitude from McCarren Field, was a recipe for disaster. It is possible – just – that the alleged overweight of the aeroplane may have taken the edge off its climb-rate and that with a lighter loading it might have cleared that cliff and other potential obstacles but, the aeroplane should not have been on that course anyway.

TWA's copy of the flight-plan shows the course from Las Vegas (LV/McCarren), to the next way-point, as 218 degrees magnetic. After an allowance was made for wind-drift, the actual course was written as 223 degrees magnetic. According to the findings of the US Civil Aeronautics Board (CAB) investigation into the accident, these figures were an error on the part of the co-pilot, whose task it was to fill-out the flight-plan document. This particular document was not counter-signed by the captain – which was an omission on his part, too. If the aeroplane had taken off from Boulder City airport, 218 degrees would have been correct – but it was incorrect for a departure from Las Vegas. Furthermore, the Las Vegas radio range was functioning correctly and its centre-line heading was 205 degrees magnetic. The crash impact point was 6.7 miles to the right of the LV range centre-line. Only one local visual beacon. 'Arden 24' was illuminated. It was located 20 miles out from Las Vegas; 2.5 miles to the right of the LV range centre-line and would normally be passed on its left-hand side.

From a regulatory and certification point of view, the inquiry found nothing wrong with the aeroplane, the pilots or the flight. Captain Williams was a highly experienced pilot with 12,024 flying hours to his name and co-pilot Gillette had 1,330 flying hours. Despite this, they were both more accustomed to routing through Boulder City than Las Vegas. Williams had flown just twice into McCarran and Gillette only six times on that route and it is believed they had made a departure from McCarran towards Burbank only once each.

According to the CAB findings, the probable cause of the accident was a failure of the captain after take-off from Las Vegas to follow the proper course by making use of the [radio range] navigation facilities available to him. Contributing factors were: 1. The use of an erroneous compass course. 2. The blackout of most of the [illuminated] beacons in the neighbourhood of the accident made necessary by the war emergency. 3. Failure of the pilot to comply with TWA directive requiring pilots to confine their flight movements to the actual on-course signals [of the relevant radio-navigation beacons known as 'radio ranges'].

There were two pilots on the flight deck that night – but which one was actually flying the aeroplane? During investigations by the CAB; House of Representatives; the FBI and the USAAC, many theories including suggestions of sabotage, Unidentified Flying Objects (UFO) or fights on board abounded, but the conclusion was that these tales were unproven.

In a quiet private ceremony led by her distraught husband Clark Gable, Carole Lombard Gable, aged 33 and her mother Elizabeth, were laid to rest on 21 January 1942 in the Mausoleum of Forest Lawn cemetery, Los Angeles. Carole's legacy is assured through her films, but her magnetic presence on screen reminds us of what a special talent had been lost.

Chapter 7

Leslie Howard

As conspiracy theories go, those surrounding the death of the British film star Leslie Howard, are right up there with the best of them.

On 1 June 1943 eight Junkers Ju 88s – twin-engine, long-range, heavy fighters – from the Luftwaffe's 14 *Staffel* of *KampfGruppe 40* (*14/KG 40*) took off at 10.00 from their airfield near Bordeaux to search for a pair of German U-boats transiting the Bay of Biscay while returning from patrols in the North Atlantic. The Bay provided rich pickings for RAF anti-submarine patrols and if either U-boat was found, the Ju 88s were to escort it back to port and protect it from possible attack by RAF aircraft. Due to poor weather conditions over the sea, the search for the U-boats was called off, but the Ju 88s continued to patrol the area which, in view of the ever-present threat from RAF bombers and long-range fighters, became a general search for 'targets of opportunity.'

Earlier that morning, at 07.35 GMT, Flight 777-A took off from Portela de Sacavém airfield, Lisbon, Portugal, bound for RAF Whitchurch airfield near Bristol in England. The aircraft was a civilian KLM (Royal Dutch Airlines) Douglas DC-3, coded G-AGBB (formerly PH-ALI in Dutch service) and named *IBIS*. The pilot in charge of the all-Dutch crew was Captain Quirinus Tepas OBE and he had with him: first officer Captain Dirk de Koning; radio officer: Cornelis van Brugge and flight engineer: Engbertus Rosevink. Among the seventeen passengers and crew, the person with the highest profile was the famous British film-star Leslie Howard – or to give him the name he was born with: Leslie Howard Steiner – he changed his name later.

When the Germans invaded Holland in May 1940, *IBIS* was hurriedly serviced and her Dutch crew flew her to England, landing at Shoreham airfield, near Brighton, on 13 May. *IBIS* was one of four KLM DC-3s that escaped to England and re-registered. Painted in standard RAF camouflage – with their civilian code letters displayed, but no RAF roundels – their crews continued to operate passenger-carrying services, but now under the banner of the British Overseas Airways Corporation. These unarmed DC-3s had been flying regular services to Gibraltar and neutral Portugal from Whitchurch airfield since September 1940 but, only on two previous occasions was the *Luftwaffe* encountered *en route*. On both occasions the airliner involved – which, coincidentally, was G-AGBB – had escaped and although sustaining damage from enemy gunfire, no person on board was harmed. By the end of the second year of these operations, over 4,000 passengers had been carried and by June 1943, four return trips per week were being flown to Lisbon.

On the morning of 1 June 1943, RAF Whitchurch received a radio message that *IBIS* had left Lisbon and a regular exchange of signals took place up to the point where at 10.54 GMT the DC-3 reported first that it was being followed and then that it was being fired upon. That was the last time anything was heard from the aeroplane.

Leading a formation of Ju 88s that day was *Oberleutnant* Herbert Hintze, flying as *Staffel Führer* and many years later he recalled the incident for the eminent aviation historian, Chris Goss.[1]

'The DC-3 was flying on a reciprocal course towards our formation. I first saw her as a grey silhouette from a range of 2–3,000 metres. Bellstedt radioed: "Indians at 11 o'clock. Attack! Attack!" I could not see markings but, considering the silhouette and construction, it was an enemy aircraft.

Oberleutnant Albrecht Bellstedt and his wingman *Leutnant* Max Wittmer-Eigenbrot were flying as the top-cover pair, above the rest of the formation and dived on the target from above. The rest of us intended attacking from below but, by the time we got into range, Bellstedt had fired and set the port engine and wing ablaze. As we came closer, I could see the aircraft was a DC-3 and had civil markings. I immediately ordered cease fire. I saw what appeared to be three parachutes emerge from the aircraft, which did not open as they were burning. The 'plane went into a flat turn and ditched, floating for a short while before sinking. There were no signs of survivors. On our return to base, we were told we had shot down a civilian aircraft with VIPs on board. I can still remember quite clearly that we were all rather angry, particularly because no one had told us, before, that there was a scheduled flight between Lisbon and the UK. If we had, it would have been an easy thing for us to escort the DC-3 to Bordeaux.'

There is conjecture to this day as to whether *Luftwaffe* High Command deliberately sanctioned the shooting down of this DC-3. Rumours of Churchill and other high-ranking officials being aboard the aircraft circulated but, since such information was tacitly, if not freely available in wartime Lisbon, the German intelligence network would almost certainly have been able to confirm or deny them. While several aboard the air-liner were prominent in their professions and active in the Allied war effort, no single person on the all-British passenger list stands out as crucial – in the grand scheme of things – to the conduct of the war.

The main conspiracy theories put forward as to why this was a deliberate ambush are:

1. A case of mistaken identity which suggested to German agents watching at Lisbon airport that Winston Churchill and his personal bodyguard Detective Inspector Walter Thompson, had boarded the aeroplane to fly to England.
2. The aeroplane was full of British agents, one of whom was Leslie Howard and this was an opportunity for the Nazis to dispose of 'a nest of vipers.'
3. Leslie Howard was specifically targeted for his high-profile presence and value to the British propaganda machine or, more specifically, Joseph Goebbels, Reichs Minister for Propaganda – according to unsubstantiated rumours, having been ridiculed in a film – had a personal desire to have him eliminated.

The British government had a strict policy that passengers on these flights were to be only diplomats, civil servants, official couriers, military personnel (including some POW escapees or evaders) and others deemed as VIPs travelling with government permission. Until this particular incident, these flights were made during daylight hours and at a cruising altitude of about 3,000 feet at which altitude the DC-3 still had sufficient range to take a course well out to sea. As a result of this latest attack, future flights on this route were – belatedly – made at night and the course flown was moved even further west, turning the journey in to a more than five-hour trip.

Almost as many conspiracy theories are attributed to the passengers as those about the prime 'target' Leslie Howard and on this fateful day, the passengers were as follows. Tyrell Mildmay Shervington was a director in the Lisbon office of Shell Mex & BP Oil Company who was flying to England for family reasons. It is suggested in some accounts of this episode, that he was an agent in the Iberian branch of the Special Operations Executive (SOE), a British espionage organisation. Wilfred Jacob Berthold Israel, an Anglo-German Jewish activist, having been instrumental in saving many lives of potential Holocaust victims through the pre-war *Kindertransport* scheme, was now rumoured to be compiling a dossier about concentration camps. He was returning to England after participating in secret talks in Madrid and Lisbon that might lead to Jews allowed emigration to Palestine. Alfred Tregear Chenhalls was an accountant in production aspects of the film industry and Leslie Howard's business manager, friend and travelling companion. Kenneth Stonehouse was a British journalist in Reuters' Washington office, accompanied on this flight by his wife, Evelyn. Ivan James Sharp was a mining engineer and an official at the UK Commercial Corporation (UKCC). He was returning from a tour of Spanish and Portuguese mines to report upon his attempts to buy up, as much as he possibly could, the Iberian supply of the mineral wolfram – also known as tungsten – a vital metal, prized by all the belligerent nations. Gordon Thompson MacLean worked for the Foreign Office as Inspector-General of British Embassies and Consulates abroad. Rumours suggest he had persuaded Frank Foley, a former British agent of MI6 in Berlin and a person credited with helping many Jews to escape from Germany before the war, to give up his seat so that McLean could chat to his favourite actor. Francis German Cowlrick was the long-serving vice-chairman of Spanish Babcock & Wilcox Company bringing his vast civil- and mechanical-engineering expertise to the British company's Iberian operations. He was returning to England on business. Mrs Cecilia Amelia Falla Paton, according to information on the Trafford (Manchester) War Memorial Selznick site, was the wife of James Paton of 53 Ullswater Road, Flixton, Lancs and as a civil servant she was on her way to take up a secretarial post at a consulate in England. Having been in the USA as refugees, Petra Hutcheon, aged 11, Carola Hutcheon, aged 18 months and their mother, Mrs Rotha Violet Lettie Hutcheon, had travelled by ship from the USA to Lisbon and were now flying to England to re-join their husband/father, a Lieutenant-Colonel in the Royal Artillery.

And then there is our final passenger, the actor Leslie Howard. Howard was an internationally-known film, stage and radio star, best known for playing the role of the disillusioned, southern gentleman Ashley Wilkes in one of the all-time Hollywood classics: *Gone With The Wind*[2].

Leslie was born in Forest Hill, London on 3 April 1893. His father Ferdinand Steiner was a Hungarian Jewish immigrant and his mother Lilian Howard, neé Blumberg, an Englishwoman of German Jewish descent, the daughter of a London barrister. Raised in Vienna while his father was employed in Austria, the family returned to London when Ferdinand joined a stockbroking company in the City of London. By that time his father had anglicized his name to Frank Stainer and Leslie had gained a sister Doris and a brother, Arthur – and later there came another sister, Irene.

Upon leaving school, Leslie took a job as a junior clerk with a London steamship company for a short time – retaining his keen leisure interest in amateur dramatics – before, with a little help from his father, finding a post as a bank-clerk in the city until the outbreak of the First World War.

At the age of 21, Leslie volunteered for the Army, obtaining a commission as a second lieutenant with the 20th Hussars. In March 1916, he married Ruth Evelyn Martin and was posted to France shortly afterwards. Having fought in the Somme campaign, he was invalided out of the army in 1917 suffering from what was then called 'shell shock.' Upon recovery, he took to the stage and after his first stage role in 1918, he soon became moderately famous in the West End theatres of London and on Broadway, New York, often portraying what the English liked to convey as 'the perfect English gentleman' – variously mixing the elements of: tall, slim in stature, intellectual, sensitive, foppish or sometimes absent-minded, to whatever degree his current role demanded. Around this time, Leslie, his wife and their son Ronald took up residence with his father and mother in their large house in West Kensington.

Leslie's son Ronald (1918-1996) was born while the family lived in South Norwood and they also had a daughter, Leslie Ruth, born in New York in 1924.

Leslie Howard is credited with appearing in nine films from the 'silent' era, making his first film appearance back in 1914 as an uncredited extra in a crowd scene in *The Heroine of Mons*.

Regarded as handsome and already well-established for a decade on the Broadway stage, the coming of 'talkies' films in the late-1920s, was his passport into the Hollywood movie scene, where there was a shortage of stage-trained voices that were essential in those early days of poor-quality sound. His English-public-school upbringing stood him in good stead. It was a rare success for an Englishman to get a foot in the door of Hollywood and he felt it would do no harm to his prospects by adopting one of his mother's given names of Howard as his own surname. Having changed his name by Deed Poll[3], as Leslie Howard he was now set for stardom.

The first of his twenty-seven 'talkies' – films he made both in the UK and Hollywood – was *Outward Bound*, starring alongside the Hollywood legend Douglas Fairbanks Junior in this Warner Brothers 1930 movie. He brought several of his stage roles to the silver screen, for example in *The Petrified Forest*,[4] in which he starred with another Hollywood legend, Bette Davis. It was during the stage production of the play that Leslie forged a lasting friendship with the, then, little-known actor Humphrey Bogart – a future Hollywood legend himself. One of his early 'talkie' international successes came in 1933, with *Berkeley Square*, a time-travelling fantasy drama in which he starred with Heather Angel. It earned him a Best Actor in A Leading Role *Oscar* nomination, but he lost out to Charles Laughton who won the award for his role in *The Private Life of Henry VIII*. Leslie's breakthrough film in British cinema came when he portrayed the foppish/heroic role of Sir Percy Blakency in the smash-hit *The Scarlet Pimpernel* (1934) alongside the Hollywood beauty Merle Oberon.

Leslie was nominated again for a Best Actor *Oscar* for his role as Professor Henry Higgins in the film *Pygmalion*[5], which he co-directed with Anthony Asquith. It was a very successful production and Higgins was a part ideally made for his talents. Leslie did not win an *Oscar*, but the film won the Best Screenplay category.

Leslie Howard's final role in Hollywood was in *Intermezzo*[6] with Ingrid Bergman in her first English-language role and Hollywood debut. Leslie was an associate producer of this film and found that was a role he rather enjoyed, too.

It is sometimes alleged that Leslie had a roving eye for the ladies. Rumours abounded after *The Scarlet Pimpernel* that Howard and Merle Oberon had developed something more than just an on-screen relationship. While in Madrid, during his Iberian tour just prior to his death, he met up with Conchita Montenegro, his co-star from the film *Never The Twain Shall Meet*[7]. Conchita's husband was a friend of General Franco and some accounts say she was merely a pawn in Howard's 'hidden agenda' to lobby Franco on behalf of Great Britain, in an effort to keep him out of the war. Another of his *femmes fatale* who turned up in Madrid was Baroness von Podewils, who was reputed to be a German agent. Although married, even back in England Leslie conducted a discreet affair, too, with an attractive blonde Frenchwoman, Violette Cunnington, who had been a production assistant on the film *Pygmalion* but later worked as his own personal assistant. Acting under the stage name of Suzanne Clair, she had small parts in *Pimpernel Smith* and *The First of the Few*. In his own last months of life, he found comfort in the arms of another *amoure*, Josette Ronserail, a Free French patriot, rumoured (again) to be working for the SOE in London. All very Mata Hari-ish!

Happily working in Hollywood, when the prospect of war with Germany looked certain Leslie bought himself out of his contract and returned to England, immediately offering his services to the government. During the Second World War, Howard became a vociferous exponent of British anti-Nazi propaganda in film, radio and in print. British and American newspapers published many of his articles about the progress of the war and he made morale-boosting radio broadcasts for the BBC. Furthermore, having had substantial experience of the dominance of Hollywood in cinema, his return to Britain coincided with his intention of stimulating his own career on this side of the Atlantic and reinvigorating the ailing British film industry at the same time. He would now launch himself into the production, direction and script-writing side of the camera as well as playing leading men in front of it. He made his home in a Tudor-period six-bedroom Grade-II listed farmhouse called Stowe Maries, in Westcott, Surrey, which he bought in the 1930s at the height of his fame and where he entertained many Hollywood stars.

At least three of his most recent films, such as: *Pimpernel Smith* (1941)[8]; *49th Parallel*[9] and *The First of The Few* (1942), all of which he himself produced and directed, had strong anti-Nazi propaganda themes in their plot-lines and it is thought these latter roles alone were enough to make him 'a marked man' to the Nazi hierarchy. In particular, the *Pimpernel Smith* film (his first as a solo director) – in which he, as Professor Horatio Smith, updated the role of the Scarlet Pimpernel to carry out the rescue of key people from the clutches of the Germans – did not pull any punches in ridiculing the Nazi party – particularly Joseph Goebbels, one of the party's most prominent leaders. The film *49th Parallel* was a joint British/Canadian production in which Leslie co-starred with some of the greats of the silver screen: Laurence Olivier, Anton Walbrook, Raymond Massey, Glynis Johns and Eric Portman. One of his finest achievements, *The First of The Few*, tells the story of the inventor of the Spitfire aeroplane, R.J. Mitchell. It was pure flag-waving stuff and Leslie's kindly portrayal of the often – in reality – blunt and irascible Mitchell in no way distorted the underlying nationalistic purpose of the story-line. It was the highest-earning British film of 1942. It is these propagandist films, broadcasts and writings that, in the main, have led to many of the stories that purport to explain why Leslie was the victim of a targeted plot by the Nazi hierarchy to assassinate him.

None of the theories, however, have been substantiated beyond reasonable doubt and his supposed involvement in matters of espionage can equally have been exaggerated by a desire to construct a neat explanation, based possibly more on screenplay-licence than on hard documented facts.

The reasons for Howard's visit to neutral Spain and Portugal appear to be to publicise these recent films, which were being widely shown in Iberian cinemas and to promote the Allied cause alongside them. He had expressed doubts to his agent and friends about the value of this tour but was swayed by an exchange of correspondence with the Foreign Secretary, Anthony Eden. In Lisbon, having completed his engagements, Howard became anxious to return to England as quickly as possible. He turned up at Lisbon airport around 5.00 pm, in the evening of 31 May, where he applied for air tickets for himself and his companion Chenhalls, on the first available flight. Parked out on the tarmac, G-AGBB was fully booked, but Leslie Howard's status was used to obtain two seats on that aeroplane. This was accomplished by two people – Derek Partridge, the seven-year-old son of a British diplomat and Derek's nanny, Dora Rove – being informed that it was no longer possible for them to make that flight.

According to an account of this incident written by aviation historian Chris Goss, the DC-3 went down roughly 200 miles north of Cape Ortegal, on the north-west tip of Spain. There has been conjecture that this area was somewhat further west than the Ju 88s of *KG 40* usually operated; giving additional traction to the theory of a premeditated interception. This location is, however, not considered sufficiently unusual to support the notion of a pre-arranged attempt to intercept and destroy that DC-3.

Furthermore, it was not the first time that a civilian aeroplane, *en route* from Whitchurch to Lisbon, had been attacked in the Biscay area. Indeed, it was G-AGBB that was intercepted by long-range fighters of the Luftwaffe during two earlier flights to Lisbon. On 15 November 1942 and again on 19 April 1943, she was attacked and damaged by gunfire but the pilots – Captains Theo Verhoeven and Dirk Parmentier – escaped on both occasions by skilful flying.

Just as on the high seas and on land, in what had, by mid-1943, become a highly-contested air-war zone, transport aeroplanes were legitimate targets and therefore interception over Biscay by the enemy was a hazard of the job that these aeroplanes and crews were employed to do – and the passengers knew the risks, too. To Britain and the Allies, Leslie Howard, an actor recognisable to audiences all over the world, was undoubtedly a great asset as a contributor to the nation's propaganda war effort. His passing, though, would certainly not be mourned by the Nazi hierarchy, to whom he might be considered as a thorn in their side. It is felt, however, that there is insufficient evidence to deduce that the Nazi regime would or could, set up an assassination operation based around the interception of a lone aeroplane, flying several hundred miles out to sea, in potentially unpredictable weather conditions and then not only to find it but also to be sure of shooting it down with complete loss of life of all on board. Rather than focussing on one incident, this interception should be considered as one of the consequences of the overall Battle of Biscay. It was reasonable for the Luftwaffe to take every opportunity to disrupt or even put a stop to what had become a frequent enemy passenger service across the Bay. Interceptions resulting in fear, damage or worse, to these transport aircraft – civilian or military – and their passengers, could only help that cause and both sides knew the score. It seems more plausible to conclude that DC-3, G-AGBB had the misfortune to be in the wrong place at the wrong time and in the absence of orders to the contrary, *Oberleutnant* Albrecht Bellstedt, was doing his duty by 'firing first and asking questions later.'

Memorial plaques to loss of DC-3, G-AGBB can be found at Lisbon airport; Whitchurch airfield and on the barren cliffs near Cedeira, a point on the Spanish coast closest to the site of the incident. Leslie Howard's personal legacy can be found in the continuing enjoyment of his catalogue of movies.

Chapter 8

Glenn Miller

15 December 1944

Anxious to get to Paris, where he was to meet up with his US Air Force Band, Major Glenn Miller accepted an innocuous invitation to fly there in an USAAF C-64A Norseman, single-engine aeroplane. It was a trip that ended his life. With its leader posted as missing, his band played on and Miller's music became one of the most enduring and instantly recognisable, sounds of the modern world. So, who was this charismatic genius, whose interpretive musical skill entranced nations?

At 13 years of age, a scruffy little kid from Clarinda, Iowa saved up money to buy himself an equally scruffy trombone. He taught himself to play and grew up to make magical music, recognisable to countless generations decades after his mysterious disappearance at the height of his fame. That kid was Alton Glen Miller and his style of music became an international phenomenon that will endure forever.

Born on 1 March 1904, Alton Glen (later changed to Glenn) Miller was the second of four children of Lewis Elmer and Mattie Lou Miller, of Clarinda, Iowa, in the USA. His father was an itinerant carpenter who moved around Nebraska and Missouri, before settling down in Fort Morgan, Colorado. In 1918, Glenn went to High School and tried his hand at playing his beat-up trombone in his Sunday school and the local town bands. By 1921, he was sufficiently confident to skip his graduation and go after a job playing with a band in Laramie, Wyoming, but which – for all his enthusiasm – failed to materialise. In those early days, after High School he moved around a few bands before playing trombone in a band run by Holly Moyer in late-1922. He got on well with the father-figure of Moyer, who helped him with tips on musical and personal style and presentation that Glenn would remember and use, in years to come.

In 1923 Glenn continued his music studies at Colorado University in Boulder, but was not a keen student; preferring to pursue an itinerant life as a trombone sideman trying to earn a living. He had already decided he wanted to become a professional musician and his enthusiasm lay in trying to improve his own playing technique and develop more experience as a would-be arranger of jazz and popular dance music. Glenn dropped his studies in 1924, but did, while at Colorado Uni, find the other love of his life – Helen Burger. He must have been smitten because he stayed in contact with Helen even though he roamed the country chasing those musical dreams. Indeed, Glenn and Helen's long-distance relationship stood the test of time and they married on 6 October 1928 in New York.

Glenn settled down in the Tenafly suburb of New York with Helen, who encouraged him to take up steady work in recording studios, on the radio and in productions on Broadway. Thus, the years from 1928 to the mid-1930s saw him working with many American big-name jazz and swing bands of the day. His contemporaries read like a directory of the golden age of jazz and swing, such as: Bing Crosby; Jimmy Dorsey; Tommy Dorsey; George 'Pee Wee' Ervin; Benny Goodman; Harry James; Gene Krupa; Ray McKinley; Ray Noble; Charlie Spivak, Jack Teagarden *et al*. Right there in the middle of them was Glenn Miller, playing, composing and arranging with and for them all.

After forming his own band, it toured major cities, recording for major labels such as Decca and Brunswick – but running a band was hard work, stressful and expensive. Despite a loan from Tommy Dorsey, Glenn's agent did not find him the scale of work required to maintain the band and he had no option but to wind it up. To cap it all, Helen fell ill and went into hospital, requiring a serious operation, the outcome of which meant she was no longer able to conceive. Nevertheless, their parental instincts remained undimmed and they were able to adopt two children, a baby boy Steven in 1943 followed by a baby girl, Jonnie in 1944.

By the end of the summer of 1938, he had found some reliable finance and a Glenn Miller Band was re-born under a vibrant new management agency. The new band recorded its first number, *My Reverie*, on the RCA-Victor Bluebird label, which reached number 11 in the US *Billboard* chart.

It has to be said that, professionally, Glenn was not the easiest person to get on with. He sought perfection for himself and expected his colleagues to focus on the same thing. Over the years, many people have tried to define the 'Glenn Miller Sound.' In very simple terms it seems to have evolved from Glenn's response to two personnel changes that were forced upon him. Now running his own band, he wanted to work with a line-up that included a tenor-led saxophone section, but also one that used the influence of the clarinet. Always with an eye on his finances and keen to maximise the work output from a new lead clarinettist he had hired, Glenn told his new musician that, when he was not doing soloes, he was to play along with the tenor sax. The significance of this was that both instruments played in the same key, but the clarinet's sound, while blending with the melody, was in a higher register. These two were then supported by harmony from the other saxophones. Furthermore, when one of his high-end trumpet-players left the band, Glenn had another clarinettist play that part. When these two novelties combined, the effect on the sound the band produced was quite startling – and refreshingly different. Audiences loved it, too and adding singers like the ebullient, seventeen-years old Marian Hutton; smooth crooner Ray Eberle and sax-man/singer Tex Beneke, Glenn's opportunity to hit the big-time finally arrived.

After a sensational impact at the Paradise Restaurant in New York during 1938, it was through 1939 that Glenn Miller's fortunes took that upward leap from which he never looked back. His radio exposure was boosted later in the year when the commercial radio company Columbia Broadcasting System – CBS – signed-up the Miller band to appear on a prime-time programme sponsored by Chesterfield Cigarettes: *The Chesterfield Show*. There was no finer accolade for musical success in those days than the *Billboard* chart and in 1939 the Glenn Miller Band notched up no less than seven number-1 hits, including the classic instrumental *In the Mood*, which went to number-1 for 13 weeks and stayed in the chart for twenty-eight weeks.

During 1940, Glenn had thirty-eight singles in the *Billboard* top-20 and of those, twenty-seven made it into the top-10, with the band hitting the top *Billboard* spot with *Tuxedo Junction* – nine weeks at number 1 and seventeen weeks in the charts.

Glenn was almost inevitably led into the movie business as the next stage of riding the crest of his wave of popularity and as a publicity vehicle for the Miller Orchestra. Its first film: *Sun Valley Serenade* can be seen as a huge success. Shot in black & white; starring Sonja Henie and John Payne, *Sun Valley Serenade* is a musical romance set in the ski resort of Sun Valley in Ketchum, Idaho and features Glenn Miller in the role of band-leader 'Phil Corey.' When it came to the orchestra numbers, it was full-on Miller-style at its magnificent best – and Glenn is reputed to have received a fee of $100,000 for his work on the film (≈ $2 million in 2024).

With the Modernaires and Tex Beneke doing the vocals on the recording, *Chattanooga Choo Choo* discs were selling like hot cakes and in February 1942, it became the first recording to have its one-million-plus sales recognised by a Gold Disc.

Between April and June, the Miller orchestra was back at work on a film set making their second movie: *Orchestra Wives* (20th Century Fox; 1942). During 1942, *Orchestra Wives* was Twentieth Century Fox's thirteenth highest-earning film, grossing $1.3 million at the box-office (≈ $24.6 million, or £18.5 million in 2024), while *Sun Valley Serenade* is believed to have made around $2.3 million[1]. During that summer, too, the Miller orchestra continued to cut discs, with ten of his seventeen recordings reaching the top-ten and *String of Pearls*; *Kalamazoo* and *Moonlight Cocktail* each capturing the top spot in the charts for several weeks.

At the very top of his game, Glenn was concerned about the war and was acutely aware that in his unique position in the minds of his country-men and -women, he ought to be seen to do his bit for the Allied war effort. Glenn's orchestra had been used on many occasions to front the great War Bond sales that were touched upon in the chapter about Carole Lombard, helping to raise over $4 million in just two War Bond rallies in 1943. At the age of 38 – currently outside the Military Draft limit – while working in Washington, Glenn Miller settled-up his business and personal affairs; dis-banded his orchestra and was inducted as a captain into the US Army Specialist Service Corps by September 1942. It fell to the United States Army Air Force (USAAF), though, to recognise Glenn's potential to help lift personnel morale through the medium of music. Furthermore, it was felt that through the medium of music, Miller would contribute to AAF recruitment by forging a link between the rapidly expanding wartime Air Force and the American public, through his huge personal following in both sectors. Having transferred to the AAF, early in 1943 Glenn was posted as Director of Bands (Training) in the Army Air Forces Technical Training Command, where he was given *carte blanche* to re-organise and modernise bands and military music for the Air Force.

The Allied Expeditionary Forces Programme (AEFP) went on air on 7 June 1944. Now, Supreme Commander General Eisenhower ('Ike') wanted the best for 'his boys' and he asked the head of the AAF, General Henry 'Hap' Arnold, to ship Glenn Miller, his orchestra, its associated radio production unit; its executive unit run by Glenn's right-hand man, Lieutenant Don Haynes – Glenn's administrative [exec] officer and his former manager in civilian-life; plus assorted staff, over to England. Popularly known as the Glenn Miller AAF Band, officially it became the American Band of the Allied Expeditionary Forces and comprising the cream of dance band and orchestral musicians, was a prime attraction on the new AEFP network.

Glenn flew to England; arriving in London on 18 June 1944, while the rest of his party of sixty and equipment sailed on the RMS *Queen Elizabeth*, which docked in Glasgow on 29 June. They, too, travelled to London by train, then re-located 50 miles north to Bedford three days later. With its personnel billeted in and around Bedford, the band was administered from nearby Milton Ernest Hall, home to the US 8th Air Force Service Command (AFSC) HQ.

In addition to recording and making live radio broadcasts, an essential part of the orchestra's duty was to stage concerts at USAAF operational stations and other, mainly military, venues. Between July and November 1944, Glenn Miller and the AAF Band played seventy-one concerts all over the UK.

Based at Milton Ernest, it was convenient for the AAF band to use Bedford Corn Exchange for live and recorded concerts, because the building was already 'wired up' by the BBC for broadcasting. David Niven, the British film star, now enters the story. After service with infantry regiments and the Commandos Niven, by now a Lieutenant Colonel, was transferred in 1944 to General Eisenhower's Supreme Headquarters Allied Expeditionary Force (SHAEF). In this role, Niven worked with the BBC; ABC and CBC as AEFP's deputy executive officer of Troop Broadcasting, assisting Colonel Edward Kirby (US) to manage an organisation that found personnel and acts to entertain Allied forces both in England and after D-Day, in Europe, by radio broadcasts. It was through this work that he liaised with Glenn Miller and the Army Air Force band to arrange public appearances and broadcasts, both live and pre-recorded.

Following the liberation of Paris in late-August, Glenn pestered his superiors for permission to travel to France to perform – making it clear that he wanted his orchestra to make a live broadcast from Paris at Christmas. Lieutenant Colonel David Niven, now elevated to AEFP's Deputy Director of Troop Broadcasting, was posted to SHAEF in Paris to organise live broadcasts and recordings from facilities in that city, supported Glenn's proposal. Glenn was told that only if sufficient radio programmes could be pre-recorded to cover a proposed absence of six weeks, would the band be released to go to Paris.

On 13 November, Glenn, accompanied by David Niven, took a scheduled flight from Bovingdon airfield (Station 112) to Orly airport in Paris. Operated by the US Air Transport Command (ATC) Europe Division, Bovingdon was the main hub for non-combat air transport movements and was run along commercial airline principles. All personnel, particularly VIPs, had to have written orders or authorisation to travel out of England and seats were allocated on a priority basis. At SHAEF HQ in Versailles, they met Ike's chief of staff General Bedell Smith and Glenn's proposal to take the ABAEF band to Paris was approved.

Back in England, Glenn continued, with great discipline on the part of himself and his musicians, to address the workload pressing in from all sides. He delegated Lieutenant Don Haynes to go to Paris to sort out accommodation and logistics for the band's stay in the city. Reporting directly to Lieutenant Colonel Niven, despite several trips, Haynes seems to have not completed his task satisfactorily, due – allegedly – to his preference for too much sampling of the delights of 'gay-Paree.' His behaviour incurred Niven's displeasure and Haynes's lapse of discipline thus becomes a significant factor in the determination of Glenn's subsequent travel plans. Haynes seems to have fallen under the influence of Lieutenant Colonel Norman Baessell, a staff officer with the 8th Air Force Service Command and based at Milton Ernest – where he met Haynes and Miller. Baessell was currently working on construction projects in northern France and was required to spend a good deal of time flying to and from England. To make these trips, he used a 35th Air Depot Group Noorduyn C-64 Norseman aircraft, one of several operated by the 8th AFSC and housed at the 2nd Strategic Air Depot, a regional bomber-maintenance and repair facility at Abbots Ripton (US Station 547). It is alleged that Norman Baessell was a man who also enjoyed 'the high-life' and he, too, plays a significant part in this story.

The ABAEF was scheduled to travel to Paris on 16 December. Don Haynes was due to go earlier to finalise accommodation arrangements, but Niven refused to deal with him and ordered Glenn to travel to Paris in his place. Glenn received authorisation to travel specifically on a scheduled USATC VIP flight from Bovingdon – subject to weather conditions – on or about 14 December.

When Glenn worked in Bedford, he lived at Milton Ernest Hall, but when working in London, he had rooms at the Mount Royal Hotel. Keen to 'get his show on the road' he contacted Bovingdon for a seat to travel on 13 December, but the weather forecast was atrocious and all flights to Paris were cancelled for that day. It was just the same on 14 December. Furthermore, he was advised that even if the weather improved thereafter, it was possible that, due to the backlog, there might not be a seat for him until at least Sunday 17th.

FATAL FLIGHTS OF THE RICH AND FAMOUS

By chance, on 14 December Lieutenant Don Haynes fell into conversation at Milton Ernest Hall with his friend Baessell. It emerged that Baessell was allowed – or gave himself the authority – to arrange his own flights and because he was advised that the weather was improving, had made an independent arrangement to fly to Villacoublay (airfield A-42), near Paris on 15 December, from the nearby RAF Twinwood airfield. Learning of Glenn's current travel delay, Baessell is alleged to have told Haynes that Glenn was welcome to have a seat on his flight and that Haynes rang Glenn in London and got Baessell to put his offer directly to him. The outcome was that Glenn accepted Baessell's offer and asked Haynes to collect him. He stayed overnight at Milton Ernest Hall and next morning waited for Baessell's aeroplane to make the short hop from Alconbury (Station 102), which served as airfield for the adjoining Abbots Ripton site. In addition to the pilot: Flight Officer (US warrant officer rank) John R.S. Morgan (age 22); Norman Baessell (44) and Glenn Miller (40) would be the only passengers. Glenn made this decision completely on his own. He did not notify his superior officer nor seek to have his authority to fly amended – or, almost certainly, <u>not</u> amended – to cover making this trip as a casual passenger in this aeroplane. As a VIP, Glenn would certainly not have been given permission to fly on anything other than a scheduled ATC flight.

Early in the afternoon of Friday, 15 December 1944, Haynes drove Baessell and Miller to Twinwood, about five miles distant. The air was chilly; the weather overcast and dull, with a cloud base estimated at 2,000 feet. It was not long, though, before C-64A, serial number 44-70285, landed and taxied to a halt near the tower, with its engine ticking over. The Norseman was a sturdy, single piston-engine, high-wing monoplane, with a fixed undercarriage and a cabin for freight or up to nine passengers. It was not equipped with structural de-icing apparatus.

Time to go. Tossing their bags into the spacious cabin, the two passengers climbed in after them. Although a be-spectacled officer may have been noticed by a few people, Don Haynes was probably the only other person who knew for sure that it was Glenn Miller boarding that aeroplane.

Morgan submitted an Alconbury-Twinwood-Villacoublay flight plan to Alconbury airfield flying control but, in view of the uncertain weather, was only given clearance for the flight to be made at pilot's discretion and under contact flight rules (CFR) only. Known in the RAF as 'visual flight rules' (VFR), this meant that the pilot had to maintain sight of the ground and sea along his route – so he would be flying at a relatively low-level beneath the cloud.

There was no flying activity by the resident RAF training unit at Twinwood that day, but the tower was manned and the Norseman's take-off from runway 23 (oriented 230 degrees) was noted as 13.55 BST by a WAAF clerk on duty. Recently, there is evidence that a youthful aeroplane enthusiast standing outside his workplace on Woodley airfield near Reading, identified a C-64 as he watched it fly approximately East-South-East. A C-64 was recorded by Observer Corps personnel on duty at a post near Beachy Head, as it flew out to sea in the vicinity of Langney Point, the latter at between 14.30 and 14.45 BST. It is believed no other C-64s were airborne during this period. There is nothing to suggest that Morgan – who was instrument-rated and had flown this route several times – was not flying on the designated SHAEF airway corridors, using non-directional radio beacons (NDB) to navigate from Twinwood via Bovingdon NDB (31 miles) then Dorking NDB to the beacon at Langney Point and thence to the Dieppe or St Valery beacons. It is also believed that the sighting near Woodley indicates the pilot had extended his flight-path from Bovingdon NDB to the prominent river Thames east of Reading (17 miles), before turning on to the Dorking beacon then Langney Point (62 miles). Rather than fly via the intermediate Northolt beacon, this would have kept him clear of west London and the capital's trigger-happy anti-V1 defences. The distance and time-frame from Twinwood to Langney Point (110 statute miles in say 45 minutes, at an average airspeed of 146mph) seems to support this assumption.

Norseman 44-70285 was never seen again after crossing the English coast. Glenn Miller and his companions disappeared without trace.

At a court of inquiry held in Bedford on 20 January 1945, the cause of the disappearance and the presumed death of the three missing men could not be established beyond the broad assumptions derived, for example, from Don Haynes; personnel at RAF Twinwood and at Alconbury; the Observer Corps sighting near Langney Point, together with meteorological air and sea conditions prevailing on the day of the flight. Data on the Woodley sighting did not come to light until 2012. The inquiry concluded that the aeroplane crashed into the English Channel from low level, due to loss of control by the pilot, possibly through disorientation; or through carburettor icing or other failure, that caused the loss of engine power – or a combination of these factors. That there were no survivors suggested death within the aircraft on impact with the sea, or drowning as a result of incapacity due to injury and/or freezing in cold water that had a survivability time of about fifteen minutes.

It was not until 18 December – when the band arrived in Paris – that the enormity of this missing aeroplane dawned on the top brass at SHAEF but, to be fair, the sudden, massive German offensive in the Ardennes (aka the Battle of the Bulge) on the morning of 16 December, had rather captured their attention. On 23 December, Helen Miller received a telegram to say her husband was missing and a public announcement went out on 24 December.

In the years since the accident, many theories have been aired and are still contended vigorously and it is not proposed to debate them again here. Aided by much documentary evidence that has come to light since, these theories have been examined forensically and disproved with academic rigour by the American historian, author and acknowledged 'Miller expert,' Dennis Spragg, whose considered opinion supports the original USAAF inquiry findings. As a result of the findings, disciplinary action was taken and heads rolled at 35th ADG; 8th AFSC HQ and 2nd SAD – but it was too late.

Chapter 9

Roy Chadwick

23 August 1947

The design of the Avro Lancaster heavy bomber marked the zenith of strategic weapons employed by the Royal Air Force in its Second World War air campaign over Europe. Its creation can be attributed to one man, Roy Chadwick CBE, chief designer at A.V. Roe & Co Ltd (hereafter: Avro) but it was built by hundreds of ordinary civilian men and women; maintained by legions of RAF ground-staff and flown into battle by thousands of young aircrew.

Roy Chadwick joined the Avro Company in 1911, rising to the post of chief designer and steadfastly remaining with that company until his tragic death in an aeroplane accident in 1947. During his working life, Roy was responsible for, or involved in the design of over 200 aeroplanes for Avro, of which 35 types went into production and of which the most successful include the 504; Baby; Avian; Tutor; Anson; Manchester; Lancaster; York; Lincoln; Shackleton and Vulcan.

Born on 30 April 1893 at Farnworth near Widnes, Cheshire, to Charles and Agnes (neé Bradshaw), Roy – the eldest son – was the latest addition to a family whose male line had been associated with engineering for four generations. His father was a mechanical engineer, working at the United Alkaline Company in Widnes but, in 1898, he joined Thomas Bradford & Co and moved the family to Salford, Manchester, where Roy began his education at St Luke's church school in Weasle, a suburb of Salford. In 1904, Charles changed his job again, this time joining the Westinghouse Electrical and Manufacturing Company and moving the family – which had grown to Roy and two sisters: Doris and May and a brother: Alan – to Urmston. Here Roy, now aged eleven, attended St Clement's church school in Urmston until aged 14. During his formative years Roy became an avid designer and builder of model aeroplanes.

Leaving school in 1907 Roy, with more than a bit of influence from his father, obtained a 'premium' three-year apprenticeship as a trainee draughtsman in the design office at British Westinghouse. By 1911, having successfully completed his apprenticeship, he decided his future lay in the burgeoning field of aviation. In September that year, he approached the A.V. Roe Aircraft Company, established in 1910 in Manchester by Edwin Alliott Verdon Roe and his elder brother Humphrey, for a job.

The teenage Roy Chadwick settled in confidently at Avro's Brownfield Mills works and by 1912 had a small team of assistants working under him, producing construction drawings for early aeroplanes in the Avro '500' series. It was with the 500's improved follow-on design, the iconic Avro 504, that Roy made his own transition from draftsman to designer. Design work on the Avro 504, begun at Brownsfield Mill in December 1912, was completed at a new, larger factory in Clifton Street, early in 1913. Orders for the Avro 504 soon came from both the War Office and the Admiralty during late-1913 and the Clifton Street factory began production from the beginning of 1914.

One of the first orders was for six aircraft for the RFC. However, following successful RNAS attacks against the Zeppelin airship sheds at Dusseldorf, in September and October 1914, a top-secret plan was drawn up to launch another ambitious air attack on the Zeppelin manufacturing base at Friedrichshafen on the German shore of Lake Constance. Six Avro 504s destined for the RFC were re-allocated to the Admiralty for the raid. The first three of this batch – numbers 873, 874 and 875, together with the first production Avro 504 delivered to the RNAS earlier in November 1914 – serial number 179 – were selected as the raiding force of four aircraft. Modified at Clifton Street, they were all converted to single-seaters, with an additional fuel tank to extend their range, being installed in the passenger cockpit.

The pressing nature of the proposed attack date, together with issues relating to pilots; aeroplanes and secrecy, in turn required the Avro machines to be dismantled and crated-up as they came off the production line – untested. It was planned to transport the crated aeroplanes, together with engines, spares and bombs, to eastern France where they would be erected for the raid. Five RNAS engine-fitters and five airframe-riggers would accompany the crates. These untried aircraft were also a completely new type to the RNAS therefore, as a sensible precaution, Avro was asked to provide a technician to offer operational guidance and oversee the erection work in the field. Furthermore, since there was no such thing as standard bomb racks and release mechanism for any aeroplanes at this stage of the war, Roy Chadwick was called upon to devise a bespoke set of racks and release gear for the bombs to be carried on these 504s. The young, single Chadwick was therefore the obvious choice for this clandestine job and he jumped at the chance of challenge and adventure.

A special train carried Roy Chadwick and the four Avro 504s from Manchester to Southampton, where they were loaded in darkness on 10 November. At Le Havre, by midnight they were transferred to a special goods train and were on their way through France destined for Belfort, close to the Alsace border. By mid-day on the 11 November, the 504s and their ground crew were at a French airship base on the outskirts of Belfort – 125 miles from the target, but the closest French military installation to the German border. With men and machines hurriedly secreted into a hangar, working out of sight of prying eyes, this is where Roy Chadwick came into his own. Under his expert supervision, all four aircraft were assembled in just four hours, together with fuel and oil. The specially-designed bomb-racks, made just for these aircraft, were fitted, their release mechanisms tested and the pilots shown how to operate the release toggles. Although the airframes were still not flight-tested, the 80hp Gnome rotary engines had been used before and when fitted to the airframe, were run up inside the hangar.

Briefed about the route and target, the British RNAS pilots were told to avoid flying over Swiss territory. Perversely, the French also insisted that no maps should be carried for the first part of the flight, to avoid repercussions upon France if any of the fliers were captured. The first 80 miles of the route therefore had to be memorised. Four 20lb Cooper bombs were attached to the racks of each aeroplane and the first three Avro 504s took off a few minutes apart. However, Flight Lieutenant Cannon, in the fourth aircraft (179), could not get airborne because his engine was losing revs badly. While taxying for another take-off, the tail-skid broke and he had to abort his part in the mission.

The remaining three aircraft all found the target and attacked independently. For their daring exploits all three pilots were awarded the Distinguished Service Order (DSO) and the French *Legion d'Honneur*. Sadly, there were no awards for the only civilian involved: Roy Chadwick.

Back in the Belfort hangar, the French were anxious for the raiders to leave as quickly as possible. Roy helped to dismantle the three surviving 504s and the whole party, together with aircraft and spares, left for England by train the next day.

Roy and his team began to improve the basic 504 model with great success. There now began a long line of variants, from the 504A in 1915 through to the 504R in 1926, with many aircraft being sold to or manufactured under licence in foreign countries all over the world.

In 1916, having moved the expanding design department from Clifton Street to a new factory site called Park Works, Avro then moved some of its operations from Manchester to Hamble near Southampton. In 1917 Roy, now Avro's chief designer, moved with the design and drawing office to Hamble.

It was not all work and no play, though. In 1915, while still in Manchester, Roy met Mary Hilda Gomersall, they became engaged in 1916 and married on 8 October 1921 at St Clement's church in Urmston. Settling into a house in Peartree Green, Southampton, they had two daughters: Margaret (married name: Dove) and Rosemary (Lapham).

It was in the summer of 1919, that Roy took flying lessons – appropriately in an Avro 504K – at the Avro Flying School in Hamble. However, although there is evidence that he flew solo, a search of Royal Aeronautical Club (RAeC) aviator licences fails to show one being issued to Roy Chadwick. It is possible that this anomaly may be related to his serious flying accident in early January 1920 which, according to his biography, prompted him to 'promise [his fiancée] Mary that he would give up solo piloting.'

Roy turned his hand after the war to designing a small, inexpensive, docile aeroplane for the post-war civil market. In 1919 his idea emerged from the factory as the Avro Type 534 Baby, a low-powered (35hp) machine that met all those parameters – but also nearly killed him. After he flew solo late in 1919, Roy made a few pleasure flights from Hamble in the second Avro Baby, G-EACQ. Airborne on 13 January 1920, his biography states that Roy was inappropriately dressed for the cold weather and passed out from the cold as he made his approach to land. His aeroplane stalled and crashed into a garden near the airfield. Roy was badly injured, sustaining a broken pelvis, left leg, right arm and one kneecap and the control column made a nasty hole in his neck. He was lucky to be alive and only with top-class surgery did he recover after several months of recuperation.

Despite drastic cut-backs in military peacetime budgets, Roy Chadwick continued to turn out designs – some being speculative and others being aimed to meet Air Ministry specifications. With the business continuing to expand in Manchester, in 1929 Roy felt it was time to return to the north. He, Mary and their daughter Margaret lived in an Altrincham hotel temporarily until, in 1930, they bought a house in Hale, Cheshire. By 1936 Avro became a subsidiary of the new Hawker Siddeley company and Roy became an Avro board member.

Roy was impressed by the types of civilian twin-engine, low-wing transport types being produced in the USA that were opening-up the small airliner market. He came up with a design for the Avro 652 Anson, a new aircraft that helped to carry Avro through the 1930s. War clouds were gathering and by the middle of the decade the Air Ministry was intent on expanding the RAF. Noted for its sturdy, reliable designs, in 1934 Avro was asked by the Air Ministry to submit a design for a twin-engine landplane suitable for coastal reconnaissance work. Roy Chadwick had already begun work on the six-seater civil Anson, which he developed from his earlier Avro Ten series and Avro 642 airliners. These latter were high-wing designs which he adapted radically to form the basis for his low-wing, retractable undercarriage Anson, which was enthusiastically accepted by the military and went into production in 1935; entering RAF service in 1936.

The Air Ministry was now thinking of what bombers it would need in the event of war with Germany. Roy Chadwick and his Avro team came up with their submission: the Avro Manchester, a massive, but 'clean-looking' aeroplane with an 80-foot wingspan. It was designed around a cavernous, unimpeded 33-foot-long bomb-bay, originally intended to carry and launch anti-shipping torpedoes and capable of carrying up to 10,000lbs of bombs, including a 4,000lb weapon. Handley Page also competed with its Halifax but crucially, they designed their aircraft around the Rolls-Royce Merlin. Concern over Merlin production caused the Air Ministry to hedge its bets and place orders with both companies, but the revolutionary Vulture engines proved to be the weakness in the Avro product.

The Avro 679 Manchester I first flew on 25 July 1939 and entered service with RAF Bomber Command in November 1940. Including the Mark II variant – built without the original third dorsal fin – 199 aircraft were completed. The Manchester airframe was considered a success but the Vulture engines gave constant trouble and it was withdrawn from operational service in June 1942.

Roy was convinced his basic design was sound and came up with a radical solution: change the wing configuration to carry four RR Merlin engines. He calculated that such action would not involve any drastic structural or aerodynamic modifications, because he had already designed the Manchester to cope with high stress factors stipulated within the original Ministry specification.

In May 1940 a momentous meeting took place in Manchester between Roy and senior Air Ministry technicians despatched by Air Marshal Wilfred Freeman KCB DSO MC, the Air Council member for Development and Production, who backed the project. Roy brought to the meeting a model of the Manchester with its current twin Vulture engine centre section built as a detachable unit. In a startling demonstration he removed the centre section and replaced it with one bearing four Merlin engines. He then re-attached the original outer wing sections and explained that although the overall wing span was increased by 12 feet, the remainder of the airframe was unchanged and that performance and safety would be greatly improved. This was the game-changer. The Air Ministry team went away convinced that Roy had found a solution to the Manchester problem and Freeman authorised the Manchester production line to be changed to the four-engine version and that those engines were to be the RR Merlin.

The contract for the Avro Manchester III (aka Lancaster) was received by Avro in November 1940 and the changeover to four-engine production began with the 200th Manchester airframe, with the first two examples being scheduled for completion by July 1941. In the meantime, the Manchester Mark I, Mark Ia and Mark II were going into squadron service.

The first Avro Manchester III – now re-named Lancaster – rolled out of the Woodford factory in January 1941 and a service evaluation aircraft was delivered to A&AEE Boscombe Down on 27 January 1941; the first production Lancaster flew on 31 October 1941 and Lancasters carried out their first operational mission with 44 Squadron at RAF Waddington on 3 March 1942.

During the Lancaster's illustrious service life, Roy was called upon to work closely with Barnes Wallis, inventor of the 'bouncing' and 'earthquake' bombs. In great secrecy prior to the famous Dams raid on 16/17 May 1943, he was consulted by the Ministry of Aircraft Production (MAP) and went on to design the fuselage modifications necessary to accommodate Wallis's 9,200lb bouncing bomb, its rack; rotation and release gear and loading procedure, in the Lancaster bomb-bay. He was also closely involved with the drastic airframe modifications that enabled Wallis's 12,000lb and 22,000lb, earthquake bombs to be carried aloft by the Lancaster with great success. Roy Chadwick's contribution to the aircraft industry and to the *Upkeep* bouncing bomb project was recognised with an award of a CBE which he received at Buckingham Palace on 22 June 1943, when Wing Commander Guy Gibson also received his Victoria Cross.

As ever, Roy Chadwick, having been promoted to Avro's Technical Director in 1945 and working at the Chadderton premises, pressed on with his next ventures – the Avro Lancastrian transport, a stop-gap passenger and freight transport that was followed by the reliable Avro York. The Allies had agreed that war-time transport aircraft manufacture would be concentrated in the USA – which gave the Americans a huge advantage that would be aimed inevitably at capturing the Atlantic and other long-distance passenger routes. The York was a sterling workhorse, but not up to such competition and so Roy set out to design a completely new aircraft: the Avro Tudor. A tail-wheel design, it was the first passenger airliner in the world with a pressurised fuselage, in this case to carry just twelve passengers in comfort. Although the design was approved for the national airline, BOAC, by the Air Ministry in 1944, the Tudor project was fated to become a political football and its passage into airline service suffered accordingly.

First flown on 14 June 1945, the Tudor 1 was aimed at the Atlantic market but, after many modifications its range came out at only 3,600 miles rather than the anticipated 4,000 miles. After yet more testing during 1947, its performance still fell short of BOAC's demanding expectations – not least because it was considered incapable of operating safely across the Atlantic. BOAC, with government approval, therefore cancelled its order and Avro had to scramble for new customers from elsewhere in the world to soak up Tudors coming off the production line. Despite the litany of problems, it seems evident that BOAC management was much keener to have shiny American Douglas and Lockheed products.

Roy Chadwick was not put off by the technical difficulties and political manoeuvring and set about designing the Tudor 2 by stretching the dimensions of the Tudor 1 so that it could carry up to 60 passengers in pressurised comfort. BOAC, QUANTAS and South African Airways placed orders for the Tudor 2, with the intention of operating it on all the Commonwealth air routes. The Tudor 2 was the biggest aircraft manufactured in Britain to date. Sadly, from its first flight on 10 March 1946, testing revealed very similar aerodynamic teething-troubles to those of the Tudor 1. Perplexed by this situation, Roy and his team again implemented numerous modifications. Once again though, these reduced the Tudor 2's performance to the extent that, following tropical trials in 1946, the aeroplane was not certified for passenger operations into and out of airfields East of Calcutta and South of Nairobi. The resultant cancellation of orders – losing out to American alternatives – was a commercial disaster as well as a technical failure and something Roy took very personally.

The whole Avro team was desperately looking for an answer to the Tudor problem. G-AGSU, the prototype Tudor 2, was still the test-bed for all the changes and on 23 August 1947, Roy decided to accompany the flight test crew to see if he could get a feel for some answers. It was not an unusual move on his part, since he made a habit of flying regularly as a flight observer, acting as a 'detective' interpreting information coming from the test pilots. On this occasion he was to be accompanied by Stuart Davies, Chief Designer and Sir Roy Dobson, Avro's Managing Director.

Some maintenance work was carried out overnight to G-AGSU during which, apparently, the aileron operating cables needed to be dis-connected and re-connected, but this latter work was not recorded and the crew was not advised. The two senior test pilots, Sidney Albert 'Bill' Thorn and Jimmy Orrell, flipped a coin for who would fly the sortie on Saturday and Thorn won the toss. Saturday was a sunny day and the three managers boarded G-AGSU with the test crew. Just before the engines were started, someone rushed out to the aeroplane to tell Sir Roy Dobson that there was an urgent telephone call for him. Saying he would not be long, Dobson left the aeroplane and walked to the control office to take the call. When he did not return soon, Roy Chadwick instructed that the cabin door be closed and the flight to get under way without Sir Roy. Chief test pilot Sidney Thorn fired up the engines and taxied out to Woodford's runway 25, where he checked the control column movement was free in all axes and opened the throttles for take-off.

In a slight cross-wind, Thorn felt the aeroplane yaw to the left and instinctively compensated by applying some opposite aileron. When that did not level the wings he applied yet more aileron, still with no visible effect – in fact his action only seemed to make the wing-drop worse. At 80 feet up, in a matter of seconds the left wing-tip dropped so much that it scraped the runway, slewing the aeroplane off course. Thorn chopped the throttles immediately, but the aircraft's momentum took it across a grassy area, ploughing at speed into trees on Shird Fold Farm, Adlington, just outside the airfield perimeter. The wings were stripped off and the forward fuselage split open, flinging Roy Chadwick out a distance of 60 yards. His head hit a tree trunk and he died instantly from the impact. The detached cockpit area came to rest in a large pond surrounded by the trees. Injured and rendered unconscious, Thorn and his co-pilot Squadron Leader David James 'Baikie' Wilson DSO DFC*, Chief of the Avro Test Section, were immersed beneath the water and drowned. Radio officer, John Webster, also died from a fatal head injury. Miraculously, two people survived; Stuart Davies staggered from the wreckage unaided with blood streaming from injuries to his face and flight engineer Eddie Talbot was rescued from the wreckage suffering from multiple injuries, from which he later recovered.

Nine months later, Air Commodore Vernon Brown, Chief Inspector of Accidents, published his report on the Tudor crash on 3 June 1948. The cause was stated as crossed aileron controls. This, it was alleged, was attributed to the actions of a person carrying out work in the fuselage of the aeroplane that required the disconnection of the aileron controls. The Chief Inspector concluded that:

'The person who removed the old [aileron] cables and assembled the new did not have any drawings to work to and relied entirely on memory and the work was cleared by two of the firm's inspectors. The aileron controls were assembled in such a manner that they operated in the reverse sense to normal control [column] wheel movements and that the requirements of design were not fulfilled. The issuing of M.A.P form 1090 [a certificate of safety for flight] by a member of the firm's inspection staff after major structural alterations was irregular.'

Due to the location of the cockpit being set well forward in relation to the wings, Sidney Thorn would have been unable to check visually the position of the ailerons when he turned the control column wheel to check freedom of movement prior to take-off.

Roy Chadwick, aged 54, was laid to rest in the graveyard of Christ Church, Woodford, close by the graves of his colleagues Bill Thorn and David Wilson. Many accolades were expressed for the life and work of this design genius and his immediate legacy was to be found in the Avro Shackleton and in the Avro Vulcan, which were designed and completed in accordance with Roy's outline drawings and notes, by his protégé Stuart Davies.

It is to the iconic Avro Lancaster, though, that Roy Chadwick's name will remain linked forever.

Chapter 10

Manchester United FC

6 February 1958

As in nature, the seeds of the team that would forever be known as Busby's Babes were sown long before they blossomed and two men can lay justifiable claim of responsibility for nurturing that growth. Alexander Matthew (Matt) Busby, a Scotsman who played soccer for Manchester City and Liverpool before the second world war, was recruited to the post of team manager of Manchester United in 1945. His predecessor, Walter Raymond Crickmer, placed great value on the development of young home-grown talent and with that objective in mind, in 1938 created the Manchester United Junior Athletic Club. The prevailing view among football clubs, in those days, was to adopt the more rapid expedient of buying-in new talent from other clubs as and when required, rather than running academy-style teams as nurseries for young prospects. Crickmer stayed on as United's club secretary when Busby took over and between them, they saw to it that United's Junior Club underpinned Busby's dream of a youthful side, strengthened with judicious injections of experience from outside, that could achieve and then – more importantly – sustain, success over a long period of time. Between 1951 and 1957, Manchester United bought only three players from outside the club. Success was indeed achieved but, in Munich at 15.03 on 6 February 1958 the dream of longevity was shattered.

Busby brought in former West Bromwich Albion; Swindon Town and Welsh international player, Jimmy Murphy to coach the reserve team and give particular attention to the Junior Club. Bert Whalley was influential as the chief coach and it was chief scout Joe Armstrong who spotted many of the future Babes. Inside two seasons United had captured its first trophy in over forty years: the 1948 FA Cup then, in 1952, the League Division One title. Matt Busby recognised that the time had come to crystalise his youth succession-plan and his action marked the beginning of the 'Busby Babes' era. The origin of the term is much debated but is usually attributed to a report in the *Manchester Evening News* about the debut of the first two young players in Matt Busby's new-look team.

Manchester United became eligible for the European Cup in the 1956/57 season and this marked the rise of the club, its management and its players, to the level of international status it enjoys across the world today.

United went through to the last eight of the 1957/58 European competition and were drawn against Red Star Belgrade, of the then Yugoslavia (now Serbia). The first leg was played in Manchester on 14 January 1958 and United took a 2 – 1 lead to Belgrade. On Wednesday, 5 February 1958, Manchester United was scheduled to play the second leg of their European Cup quarter-final tie against Red Star, in Belgrade, followed on the Saturday by a key League match. Involving a round-trip of 2,000 miles, the only way to meet this tight playing schedule was to travel by aeroplane. The club decided it would not rely on a scheduled airline flight but would splash-out and hire a British European Airways (BEA) aeroplane for a charter flight. It seemed an extravagant expense at the time, but they had an important league fixture against League leaders Wolverhampton Wanderers coming up the following Saturday.

The Manchester United squad taken to Belgrade numbered seventeen players. The eleven who played in the match on 5 February were: Harry Gregg, goal; Bill Foulkes; Dennis Viollet; Albert Scanlon; Ken Morgans; Bobby Charlton; Tommy Taylor; Eddie Colman; Mark Jones; Duncan Edwards; Roger Byrne (captain). The following six players travelled but did not play in the match. Ray Wood, goal; Jackie Blanchflower; Johnny Berry; Geoff Bent; Liam Whelan; David Pegg.

Thirty-nine persons (thirty-three passengers and six crew) flew on the outward trip but on the return journey from Belgrade, five additional passengers were allowed on board, making a total of forty-four persons (thirty-eight passengers and six crew) for the trip back to Manchester (see Appendix 3).

The crew for the outward and return trips was as follows. Pilot in command was Captain James Thain, an ex-RAF bomber pilot who saw active service in the Second World War. Thain left the RAF in 1946 to join BEA as an airline pilot and had been a captain since 1951. Having done a conversion course in 1955, at the date of this incident he had accumulated 1,722 hours on the Ambassador aircraft and 7,337 flying hours in total. Unusually, his co-pilot for this trip was also a BEA captain, Captain Kenneth Gordon Rayment. He, too, was a former RAF pilot in the Second World War, during which he was awarded the Distinguished Flying Cross (DFC). Rayment had a total of 8,463 flying hours, of which 3,143 were on the Ambassador. In terms of BEA service, Captain Rayment was slightly senior to Captain Thain, but his subordinate role on this flight was only an oddity brought about by the charter crewing situation and had no operational significance. It had come about when Captain Thain was asked if he would like to take the charter flight. Thain agreed and suggested to BEA operations that Captain Rayment, with whom he was friendly and who was returning to flying duties after a hernia operation, be rostered as his first officer. The radio officer was George William 'Bill' Rodgers; cabin steward was William Thomas 'Tom' Cable and the two cabin stewardesses were Rosemary Cheverton and Margaret Bellis.

The aeroplane was an Airspeed AS 57 Ambassador 2, which was assigned the class name of *Elizabethan* in BEA service. It is a high-wing monoplane with a tricycle (nosewheel) undercarriage, powered by two Bristol Centaurus 661 piston engines and this aircraft was registered G-ALZU, with the name: *Lord Burghley*. With Captain Thain in command and flying both sectors from Manchester Ringway airport to Belgrade, the outward flight via Munich was made on Monday, 3 February. The weather in Manchester was cold, with a grey overcast and fog, which delayed take-off for an hour. The weather deteriorated *en route* so that, by the time the aeroplane reached Belgrade, it was snowing hard. However, the landing was without incident and on disembarking, it was found that this flight was the only one, thus far, to make it in to the airport that day.

The match was played in the afternoon of Wednesday, 5 February at the JNA Stadium in Belgrade, in front of a crowd of 52,000. The referee was Herr Karl Kainer of Austria and Red Star fielded a team comprising: Mitic (captain); Beara (goal); Kostic; Tasic; Toplac; Sekulerac; Borozan; Popovic; Spajic; Zekovic and Tomic. The final score was 3 – 3, but the Babes had won the tie 5 – 4 on aggregate.

Next morning, Thursday 6 February, in high spirits and some heavy heads, the victorious United team was taken to Belgrade airport for the trip home to Manchester. BEA Ambassador G-ALZU waited on the tarmac and it was a happy crowd of passengers that wandered aboard. The aeroplane had insufficient range to reach Manchester in one hop, so the flight-plan allowed for a re-fuelling stop at Munich-Riem airport in Germany, a three-hour flight sector. Although it was contrary to BEA rules, by mutual agreement, Captain Thain and Captain Rayment agreed that Rayment would take the left-hand cockpit seat and fly the aircraft on both sectors back to Manchester. Thain would remain pilot-in-command but take the right-hand seat, where he could see all the necessary instruments to assume the role and duties of a first officer. As a normal precaution the aeroplane anti-icing system was used during the descent into Munich-Riem. The landing was made without any difficulty, although the cloud base was around 500 feet and it was snowing and sleeting at the time, with slush and snow across the aerodrome and on the runway. Upon landing at Munich, at 2.17 pm local time, the passengers were allowed off the aeroplane to take refreshment while re-fuelling was completed. Captains Thain and Rayment obtained weather information for the next sector, then Captain Rayment and radio officer Bill Rodgers checked in at the BEA office. William 'Bill' Black, the local BEA station engineer supervised the re-fuelling. He had to use the upper wing surface filler-access, rather than the pressure-fed, under-wing access points, but it is significant that in doing so, he stood and walked without difficulty on the wing during the filling operation.

Bill Black asked Captain Thain if he wanted the wings de-iced and Thain said he considered it un-necessary. The passengers were re-boarded and at 3.20 pm (local) the aeroplane moved away from the terminal and taxied towards the slush-covered runway. Cleared to roll, Captain Rayment opened the throttles as usual but, gathering speed, both pilots heard a change of engine note from the left engine and Thain spotted a change in boost pressure. Rayment over-rode Thain's grip on the throttles; slammed them closed and aborted the take-off. Braking brought the aeroplane to a halt on the runway. The control tower gave permission to back-track down the runway for another attempt at take-off.

Rolling again, Thain watched the boost gauges intently while Rayment pushed the throttle levers forward and he followed the movement with his own hand. Full power was not reached, though, before the left engine boost went way above limits. This run was aborted halfway down the runway and once again the aeroplane braked to a crawl. Not unreasonably, this to-ing and fro-ing caused much consternation among the passengers inside the cabin. Captain Thain received permission from Air Traffic Control to taxy back to the terminal this time. He took control of the aeroplane while Captain Rayment made an announcement to the passengers explaining what was happening, then Rayment took back control and brought the aeroplane to a halt on the terminal apron.

Engineer Bill Black was soon on the flight deck hearing about the boost surges. Black said it was a common problem with the Ambassador at [relatively] high-altitude airports such as Munich. Since the recommended procedure of opening the throttles slowly had not worked then the engines could be re-tuned overnight. Matt Busby, with Saturday's crunch game looming, had made it clear that he was keen to get back to Manchester that evening and Captain Thain echoed that view. It is quite possible to regard this moment as one of the turning points in the disaster.

They discussed opening the throttles even more slowly and agreed that, although it would lengthen the take-off roll, the runway was long enough to do this safely. De-icing was again mentioned and looking out of their respective windows, both captains felt the wing surfaces showed nothing to suggest it was necessary. There was still slush on the runway, but it was the responsibility of the airport management to clear it if they thought it necessary and BEA instruction manuals left the final decision to take-off in such conditions to the pilot-in-charge.

With the dashboard clock at 4.03 pm (local), Captain Rayment lined up the Ambassador on runway two-five. They agreed to take-off, with Thain watching the boost gauges like a hawk. Opening the throttles slowly kept the boost steady at the recommended 57½-inches. Throttles fully open; both engines at full power – and no wavering of the gauges and no warning lights. Outside the windows, slush and water were being thrown up like a bow-wave from the wheels. The speed increased, with Thain calling it out to Rayment. At 85 knots the left engine started to surge again, but Thain eased the throttle back, then gently pushed it forward. The surging stopped.

At 105 knots, Rayment eased back the control column to reduce pressure on the nose-wheel. Thain called out 117 knots – the critical speed at which they could no longer safely abort take-off (known as the 'V-1' point). Unstick speed (minimum safe flying: 'V-2') was 119 knots and when he heard that called out, Rayment would pull back on the column.

He never heard it.

Two-thirds down the runway and the undercarriage was running into undisturbed slush, the airspeed stuck at 117 knots then, alarmingly, Thain sensed deceleration as the speed fell back to 113 knots, then all the way down to 105 knots. In the aftermath, he was particularly adamant about these numbers. The nose-wheel had returned to the runway but now, still pulling on the column; altering the elevator trim-wheel; trying to coax the aircraft off the runway; Rayment called 'wheels-up' then yelled: 'Christ, we won't make it!'

The aircraft never took off.

With the throttles closed, Rayment hit the brakes but it was so close to the end of the tarmac that the impetus of the Ambassador carried it off the end of the runway, across 300 yards of run-off grass and through the wooden perimeter fence. It careered across a small road, then the left wing slammed into a house and trees on the other side of the road, tearing the wing off outboard of the engine. The left tailplane was also torn off and sprayed with aviation fuel, the house caught fire. Sliding 100 yards further, the right-side rear of the aircraft impacted a wooden garage, tearing away the fuselage to the rear of the wing. Petrol drums in the garage exploded and flames enveloped the building and this part of the fuselage. What remained of the wings and forward fuselage slid to a halt 70 yards further on. Both fuselage parts had been ripped open and crunched up by impacts with various obstacles. The nose-wheel had been retracted, but the main undercarriage and left engine were torn off and the left side of the flight deck was stove in when it hit the trees, trapping Captain Rayment inside.

In the silence that followed, it began to snow heavily.

Twenty passengers and crew died at the scene. Three passengers and crew died from their injuries, in hospital several days or weeks later. Twenty passengers and crew survived, with most of them suffering varying degrees of injury from minor to critical. To say that anyone was unhurt, though, is not strictly accurate, since all suffered knocks, bruises, abrasions and shock. Stories of heroic rescues – including the efforts of Harry Gregg – and the immediate aftermath, abound and can be read elsewhere. Harry, to all intents unhurt, was later found to have sustained a fractured skull – from which he recovered.

Perhaps the luckiest person in all this chaos was journalist Miroslav Radojčić, who wrote mainly political material for a Serbian daily newspaper *Politika*. He was, however, an ardent football fan and thought, rather than just write a match report, it would be a good idea to get a different angle on life within Manchester United. After chatting to some of the players and staff at the banquet, he went to his office and worked on his story overnight. Next morning, on the spur of the moment, he remembered Matt Busby's invitation to visit Old Trafford any time for a story. He would, he decided, take up that offer right now and join them on the aeroplane, where he could interview some of them for a feature article about United. Back in his flat, Radojčić snatched a few hours' sleep, rose somewhat hastily, packed a bag and set off for the airport. There were seats to spare; Matt Busby would let him travel, so he didn't need to worry about a ticket. All he needed was his passport and visa documents. He reached the departure gate – only to find he had forgotten to bring his passport and visa! Back home he went but, by the time he returned to the airport, the bird had flown – he had missed the flight. The story he wrote for the newspaper that day was far removed from the one he had envisaged – but he was alive to tell the tale!

Two independent inquiries into the accident were initiated by the United Kingdom government. These were both overseen by His Honour Judge Edgar Stewart Fay QC and they took place in April 1960 ('first') and April 1969 ('second'). Because the accident occurred on German soil, it was investigated by a Commission of Inquiry appointed by the Federal Republic of Germany, whose report was published on 31 January 1959. The translated German report was published in the UK by the Ministry of Transport & Civil Aviation as Report CAP 153. In its summary the German inquiry considered the accident was due to the following:

> 'During the stop of almost two hours at Munich, a rough layer of ice formed on the upper surface of the wings due to snowfall. This layer of ice considerably impaired the aerodynamic efficiency of the aircraft, had a detrimental effect on the acceleration of the aircraft during the take-off process and increased the required un-stick speed. Thus, under the conditions obtaining at the time of take-off, the aircraft was not able to attain this speed within the rolling distance available. The decisive cause of the accident lay in this.'

This pronouncement placed responsibility for the removal of snow and/or ice from the aeroplane in the hands of Captain Thain and thus, the blame for the accident. It was Captain Thain's contention, however, that uncleared slush on the runway – the responsibility of the airport authorities – retarded the aircraft's take-off velocity to the extent that it could never reach unstick speed within the runway length.

Disagreeing with the German findings, Captain Thain and BALPA, the British Airline Pilot Association, took their case to the UK Minister of Transport & Civil Aviation who appointed Judge Edgar Fay to conduct an independent review of the German findings and Captain Thain's representations. The first Fay Commission hearing was due to begin on 28 September 1959. Due to fresh evidence being presented by Thain/BALPA to the German authorities on 8 September 1959, the Fay Commission postponed its start. It was eventually re-scheduled, as a public hearing, for 4 April 1960. On 14 March 1960, the German authorities rejected the Thain-BALPA request to re-open the German investigation. The first Fay Commission therefore proceeded with its own hearing from 4 to 7 April 1960 in London, with

FATAL FLIGHTS OF THE RICH AND FAMOUS

Judge Fay as chairman, assisted by assessors Captain R.P. Wigley and Professor A.R. Collar. The terms of reference for the first Fay Commission were announced in the House of Commons and the first Fay Commission report was published on 18 August 1960 as Report CAP 167 which concluded that, in relation to Captain Thain's performance of his duty:

> 'In our opinion Captain Thain did not take sufficient steps to satisfy himself that the wings of the aircraft were free from ice and snow, but that in our opinion he did take sufficient steps to ascertain whether or not in the conditions prevailing at the time, the runway was fit for use and did take sufficient steps to ascertain the cause of the difficulties encountered on the first two attempts to take-off before making a third attempt.'

As a result of investigative work on 'slush–drag' at the Royal Aircraft Establishments (RAE) Farnborough and Bedford, new information was passed to the German government, which re-opened its investigation on 21 November 1964 and issued a second report. This was translated and published in the UK under CAP 292, but in essence the German conclusion as to the cause of the accident remained unchanged.[1]

The stigma of the German report findings spurred James Thain to continue his own investigation and in 1968 his case came to the attention of Prime Minister, Harold Wilson. Wilson was convinced that Thain was the victim of an injustice and he let it be known in government circles that he was in favour of another inquiry, particularly as a result of the vast amount of new scientific knowledge that had come to light about the effects of slush-drag in aeronautical situations. With much media attention generated by the Prime Minister's comments, the second Fay inquiry was the outcome. A key piece of evidence in the German investigation came in the form of a copy of a football fan's photograph of G-ALZU, purporting to show evidence of ice on the wings while it was parked outside the terminal. This evidence was convincingly discredited by the second Fay inquiry. Further evidence, technical and scientific, had emerged during the ten years since the accident, when little or nothing was known about the detrimental effects of slush-drag on aeroplane velocity. Experiments conducted at RAE Farnborough and Bedford also supported Captain Thain's theory.

On 10 June 1969, in the House of Commons, the Minister of State for the Board of Trade, William Rodgers MP, made the following announcement:

> 'The report of Mr Fay's [second] inquiry is being published this afternoon. Mr Fay and his colleagues were asked to consider whether blame for the Munich accident of February 1958 was to be imputed to Captain Thain. Following protracted and exhaustive examination of all available evidence which included re-examination of a number of the witnesses, the inquiry has concluded that:

1. The cause of the accident was slush on the runway.
2. It is probable but unlikely that wing icing was a contributory cause.
3. Captain Thain was not at fault with regard to runway slush.
4. Captain Thain was at fault with regard to wing icing, but, because wing icing is unlikely to have been a contributory cause of the accident, blame for the accident cannot in this respect be imputed to him.
5. Captain Thain was at fault in permitting Captain Rayment to occupy the captain's [left] seat, but this played no part in causing the accident.

Mr Fay and his assessors, in accordance with their terms of reference, have, therefore, reported that in their opinion Captain Thain cannot be blamed for the accident.

Her Majesty's Government accept this finding and the Government of the Federal German Republic has been informed.

I am sure the House would wish me to congratulate Captain Thain on the successful outcome to his long campaign.'[2]

It had taken Captain James Thain eleven years and four inquiries (two British and two German) to clear his name. He was, however, dismissed by BEA in 1960 for breaches of company regulations relating to not occupying the left-hand seat while captain-in-command and departing with some snow on the wing surface. He never flew again and died on 6 August 1975 as the result of a heart attack. Perhaps he was the 24th victim of the crash?

While Matt Busby embarked on the long road to recovery from his critical injuries, it fell to Jimmy Murphy to act as stand-in manager; to build a team to carry out its fixture commitments in the remainder of the 1957/58 season. It is a credit to Murphy; a few of the survivors and the up-and-coming youngsters of United's youth and reserve teams, that United managed to finish ninth in Division One; reach – but lose – the FA Cup final and reach – but lose – the semi-final of the European Cup, by the close of the 1957/58 season. After a long recuperation, Matt Busby returned to hands-on management of the club in the 1958/59 season, setting out once again to build a new squad worthy of the name Busby's Babes. It took several years, but Manchester United won the FA Cup in 1963; were champions of League Division One in 1965 and 1967, and in 1968 became the first English club to win the European Cup.

Chapter 11

Buddy Holly

3 February 1959

Many wrote off the future of rock-and-roll as just another passing fad – but the pessimists were confounded – it was here to stay! In 1959, though, the brave new genre of Texas rock and roll was shaken to the core by news of the untimely deaths of three of its leading exponents: Buddy Holly, Ritchie Valens and Jiles P. Richardson Jr (aka The Big Bopper) in an air crash. Their brand of rock-'n-roll music – an energetic fusion of rock-a-billy, rhythm-and-blues and country-and-western – lived on, emulated by many, thriving over the years and sounding as fresh now as it did back in 1959.

Born in Lubbock, Texas on 7 September 1936, Charles Hardin Holley was the youngest of four children of Lawrence Odell, a tailor by profession and Ella Pauline Holley; his brothers being Larry and Travis and his sister: Patricia. It was his parents who nicknamed their youngest son 'Buddy' and that was the name he used in his professional career. The family name is correctly written as Holley, but Buddy's stage name – Holly – came about when a spelling error was made on his first recording contract document and he decided to stick with that version.

At Roscoe Wilson Elementary school, Buddy befriended Bob Montgomery and by the time they reached their teenage years, Buddy and Bob were singing and strumming guitars as a duo; imitating the latest performers to be heard on the local radio station. In 1952, at Hutchinson High school, they played together as 'Buddy and Bob' on Lubbock's radio station, playing a mixture of 'blue-grass', 'rhythm-and-blues' and 'rock-a-billy' music and showcasing their own songs, reflecting Buddy's personal blend of these musical influences. Adding bassist Larry Welborn to their line-up, the trio performed at local venues, often as a warm-up act for touring 'big-name' singers or groups, such as Elvis Presley and Bill Haley and His Comets. Changing his own music style to rock-'n-roll and performing in public on stage and radio, brought Buddy to the attention of a Nashville talent scout and in January 1956, a recording contract with Decca Records followed, securing him a one-year deal with Decca at the renowned record producer Owen Bradley's Barn studio in Nashville, Tennessee. Buddy Holly was now on the cusp of a new vista of rock-'n-roll and in less than two years, he would transform the popular music scene and become one of the most influential song-writers and musicians in the evolution of the rock-'n-roll genre.

Towards the end of 1956, Decca did not renew Buddy's contract and he was taken on by Norman Petty, a musician/singer who had built a state-of-the-art recording studio in Clovis, New Mexico. Buddy re-arranged the line-up of his group to a quartet and came up with a new name for the group. 'The Crickets' group was born and its line-up became the template for the classic four-piece groups that came along thereafter. Buddy did vocals and lead guitar; Niki Sullivan played rhythm guitar; Joe B. Mauldin played bass and the drummer was Jerry Allison. When Niki Sullivan left the group late in 1957 it continued as a trio.

Buddy was keen to have a recording of *That'll be the Day* released. He, Jerry and Norman Petty re-worked the music – making it livelier and pitching it a little lower – brought in three backing vocalists and, with Larry Welborn standing-in for Joe Mauldin, on 25 February 1957 they cut a disc in Norman's studio. Released by Brunswick Records on 27 May 1957, success came slowly, although when it did, it changed the lives of Buddy and The Crickets for ever. In, too, came the sartorial elegance for which Buddy is best known – business suit or tailored jacket and slacks, white shirt and tie or bow-tie – which together with those heavy-rimmed spectacles and his Fender Stratocaster electric guitar, the package became Buddy Holly's trademark.

As well as being their record producer, Norman Petty became manager of both Buddy and of The Crickets – and he would come to acquire a detrimental influence on Buddy's finances.

Things looked up in September, when Buddy (backed by The Crickets) recorded the song: *Peggy Sue* ('B' side: *Every Day*). It became an instant success and went into the US top-10 list. This was followed up in October 1957 by the single *Oh Boy* ('B' side: *Not Fade Away*), which shot to No.10 in the US charts and to No.3 in the UK. Over just the next eighteen months, Buddy and The Crickets between them, issued around sixty singles, twenty-five of which were classed as 'hits.' Ten singles went into the US Top-100 popular music chart; eleven in the R&B chart and twenty-seven singles in the UK Top-100 chart.

When 'The Biggest Show of Stars: Fall Edition' tour began on 6 September 1957 in Pittsburgh, it ran for eighty shows across twenty-eight US states and five Canadian provinces before closing on 27 November in Richmond, VA. Buddy Holly and The Crickets were billed up there with some of the all-time great names in pop music, such as Fats Domino, The Everly Brothers, Chuck Berry, Paul Anka, Frankie Lymon and Eddie Cochran. Buddy was a rising star and on 1 December 1957, that secured him the first of two appearances on the Ed Sullivan Show on primetime TV. Shortly afterwards, *That'll be the Day* topped one-million sales.

Since he hit the big time just six months earlier, Buddy's feet had hardly touched the floor. Writing, recording, publicity appearances, touring the world in live shows made for a frantic lifestyle.

Almost the whole of March 1958 was taken up by a highly successful fifty-show, twenty-five-day tour of the UK, with their travelling show-host, the English singer/comedian Des O'Connor. After just a few days back home in the US, Buddy and The Crickets were back on the road for a mammoth forty-one-venue tour of the country with wacky disc-jockey and impresario Alan Freed's 'Big Beat Show.' On the bill with artists such as Chuck Berry, Jerry Lee Lewis, Buddy performed in 68 shows in the space of forty-four days, at venues across the mid-west and eastern states of the USA and in Canada. In May 1958, The Crickets line-up went back to four when Tommy Allsup joined and took up lead guitar while Buddy reverted to rhythm.

One example of Buddy Holly's impetuous nature occurred on 19 June 1958, while he was in New York to record *Early in the Morning*. He took the opportunity to visit his publishing manager and took an immediate shine to the receptionist Maria Elena Santiago a young lady from Puerto Rico. They agreed to a date later that day and before the evening was through, Buddy proposed marriage to her – and she had accepted. They married in Lubbock on 15 August 1958 and made their home in Greenwich Village, New York. It was around this time that Buddy and Maria became of the opinion that his recording royalty income was being manipulated by Norman Petty in such a way that Buddy was not as financially well-off as he thought.

It was financial pressure, therefore, that prompted Buddy to join yet another tour in January 1959. He was still disputing his royalties and touring income payments; Maria was pregnant and a family would bring more demands – and he was keen to set up his own recording studio, going into song publishing and record producing.

Leaving Maria at home, on 19 January, he joined 'The Winter Dance Party' tour of the US mid-west states. It set off from Chicago, where he and his backing musicians: Tommy Allsup, guitar, Waylon Jennings, bass and Carl Bunch on drums met the other acts. They shared the billing with two newcomers to rock-'n-roll: Ritchie Valens and Jiles P. Richardson, the latter known as 'The Big Bopper.' Also on the tour were Dion and the Belmonts and Frankie Sardo. Milwaukee was the first stop and it was planned to continue on the road until 15 February. In a harsh winter, it was bitterly cold when the tour set off in a tour bus which as it turned out was neither comfortable nor reliable.

In the northern states of the American mid-west, the winter of 1958/59 was one of the harshest in living memory. Boy, was it cold! Frequent snow storms covered the ground and temperatures hovered around zero degrees Fahrenheit. Whoever conjured up a rock-band road tour by bus, criss-crossing those states in the depths of January and February was ill-advised. Putting publicity-hungry stars on the road in such conditions with abysmal route planning and poor logistics has a whiff of exploitation about it. In retrospect, the 1959 Winter Dance Party was rightly dubbed the Tour from Hell.

The assembly point for the touring acts was Chicago. Buddy and his backing group travelled there from New York by train. From Chicago, a bus had been hired by the booking agency to transport the performers, their baggage and instruments to the first venue at the George Devine Ballroom in Milwaukee, Wisconsin on 23 January and thence to the other twenty-three gigs. They were expected to sleep on the bus and buy their own meals at 'comfort stops.' In another cost-cutting move, no 'roadies' were employed on the tour, so the musicians had to load, unload and set up their own equipment at all the venues. It was a logistical nightmare! So, why did Buddy take it on?

Now residing in a stylish modern apartment at 11 Fifth Avenue, New York, Buddy Holly was contacted by General Artists Corporation talent-booking agency and invited to headline this tour. He had parted from the (original) Crickets and was trying to make a name for himself as a solo artist and nurtured ambitious plans to set up his own recording studio. He needed money though. There were financial and legal hurdles relating to songs and recordings made with the Crickets; his disc royalties due from Petty were not coming through and he was taking legal action to recover what he estimated to be around $100,000 ($\approx$ over $1million in 2024[1]) and on top of all this, he had a pregnant wife to support. Personal appearances were currently his only really reliable source of income. He was not greatly enamoured by the prospect of leaving his wife behind for a long and hectic tour schedule in the middle of winter but, hell, the money was good and paid in hard cash – so he agreed to go. He needed a backing group and persuaded Waylon Jennings, Tommy Allsup and Carl Bunch to join him. The promotors venue adverts would refer to them as 'Buddy Holly and The Crickets.' Ritchie Valens and The Big Bopper were both solo acts and contracted to the same booking agency company when they, too, were asked to join the tour. The other acts on the tour were Dion (DiMucci) and the Belmonts and another up-and-coming singer called Frankie Sardo.

One show every day for three and a half weeks, non-stop; travelling hundreds of miles between venues on two-lane roads in winter, was a tough schedule. It should be remembered that these were the days before the current inter-state highways were built. It was a great consolation, therefore, to play one or sometimes two shows to packed audiences of adoring, screaming, mostly teenage fans. Buddy and his 'Crickets' also played as backing musicians for the other acts. It kept everyone's adrenaline running – which was a great deal more than could be said for the transport they used. The singers and musicians were all accommodated aboard a single bus, but the chartered tour buses did not cope well with the harsh conditions. These mostly ex-Trailways coaches, sold off to be re-used as school buses, suffered engine and heating breakdowns, requiring twelve replacement vehicles to be injected at various points *en route* at short notice and all of which were generally of poor quality.

The crunch finally came during the 330-mile drive from Duluth, MN to Appleton, WI. There was to be an afternoon show in Appleton, then an evening show in Green Bay. Leaving Duluth around midnight on 31 January, the temperature hovered around 0 degrees F (-15 degrees C). Ten miles south of Hurley, struggling up an incline on a lonely snow-bound stretch of Highway 51 near Pine Lake, WI, the bus broke down and the heater packed up. Everyone on board was totally inadequately dressed to be out in such extreme conditions and several already showed signs of colds or 'flu. After two hours, a passing trucker – who saw their predicament but didn't stop – informed the Iron County sheriff in Hurley, WI, of a stranded bus with people near the Lake. Realising the danger, the sheriff quickly rounded up a posse of cars and set off to the rescue.

The tour group was lucky; there was a real chance they could have died of exposure. Even so, Carl Bunch suffered frostbite to his feet and was admitted to Iron County Hospital, MI for treatment and recovery. Re-joining the Tour on 5 February, he would come to thank his lucky stars for that stay in hospital. Despite their narrow escape, the rest of the Party grabbed a few hours' sleep in a hotel before going on to Green Bay by Greyhound bus and train to do the late evening show on 1 February. The matinee show in Appleton was abandoned. It was at this point that Ritchie Valens, Buddy and Dion all took turns on the drums, standing-in for the absent Carl Bunch.

February 2 was supposed to be a wonderful day of rest, but the agency blew that notion by booking a show at short notice in Clear Lake, Iowa, so the Party hauled itself back on the road in a replacement bus, still having had no proper break. Another 350 miles in a noisy, re-conditioned ex-school bus brought them to the Surf Ballroom in Clear Lake, run by Carroll Anderson. The show went off very well; all the acts giving their best, despite whatever sniffles, tiredness or the debilitating effects of freezing cold rides in buses, they might be feeling. It was clear, though, from the reaction of the 1,300 strong audience – who paid $1.50 entrance – that it was the great Buddy Holly who was the 'hottest' act of the evening – he was the one they all wanted to see and hear.

The next venue was in Moorhead, MN, a 370-mile, ten-hour bus ride away. After the Green Bay debacle; missing a show; the loss of the rest day and the long haul to Clear Lake – and not least, the mind-numbing cold; sleeping upright on hard seats; breakdowns; few showers and running out of clean clothes – Buddy Holly had had enough. He was determined to hire an aeroplane and fly up to Fargo, where he promised himself he would luxuriate in a hot shower, get his laundry done and have time for a nice long sleep. He would wake up, dress clean and smell sweet; in time to do the show at the nearby Moorhead Armory.

Hearing of his intention before the show, Carroll Anderson said he could fix up an airplane for Buddy. He rang Dwyer Flying Services at Mason City municipal airport and unable to talk directly with his pal, the owner, Hubert Jerome 'Jerry' Dwyer, Anderson hired an aeroplane through one of his employees, Roger Peterson. Peterson agreed to fly Buddy and two band members to Hector Airport in Fargo, ND. Moorhead was an adjacent town just outside the Fargo city limits and over the state line in Minnesota. The price agreed was $36 for each passenger (3 x $36 = $108,

≈ \$1,160 in 2024) and Buddy accepted the deal. Later, Dwyer confirmed the hire and that Peterson a 21-years-old pilot would do the trip. Dwyer allocated a single-engine, comfortable four-seat Beechcraft 35 Bonanza, registration number N3794N for the flight, which Peterson spent the rest of the day preparing. The flight was not scheduled until after midnight, so the Bonanza was put away in a hangar until then.

Buddy offered one seat to Dion DiMucci but he turned it down as being too expensive, so Buddy invited Waylon Jennings and Tommy Allsup to fly with him and they both accepted. Musical instruments could not be carried and had to stay on the bus to meet up later. There was, however, room for personal luggage and a few parcels of dirty laundry, which Buddy undertook to have laundered for the other acts.

A series of events now took place that were literally to become matters of life and death. First, The Big Bopper was clearly feeling the effects of 'flu and in an act of sympathy and indeed kindness, Waylon Jennings gave up his place on the aeroplane to Jiles Richardson. When he learned of Buddy's proposed flight, Ritchie Valens, who had never flown before and was very keen to have a trip in an aeroplane, kept badgering Tommy Allsup to let him take his place. To put an end to his pestering, eventually Tommy agreed to toss a coin for the place. He flipped a coin, told Ritchie to call it. Ritchie called 'heads' and 'heads' it came down. Thus, Tommy Allsup gave up his seat to Ritchie Valens at the last moment. Shortly after midnight on 2/3 February, Buddy, Ritchie and Jiles Richardson Jr were driven to Mason City airport.

Dwyer Flying Services had a Federal Aviation Authority (FAA) air-carrier operating certificate with an air taxi rating that permitted the carrying of fare-paying passengers in accordance with VFR by day and night – a crucial stipulation. Roger Arthur Peterson had 711 flying hours in his logbook, of which 128 were on the Bonanza. He was currently qualified to fly in Visual Meteorological Conditions (VMC) using Visual Flight Rules (VFR) procedures only and his employer, Jerry Dwyer, was aware of his qualification status.

Nine months prior to this date, Roger had completed 52 hours training for instrument flying but, although he had passed the theory papers, he had failed the instrument flying rules (IFR) air test. It is important to note, therefore, that Roger was not currently certified to fly in instrument meteorological conditions (IMC). Essentially this is in weather or light conditions where one could not see the ground and must therefore fly by reference solely to cockpit instruments. It is also significant that Roger's IFR training was in aircraft that were fitted with a 'conventional' artificial horizon – a vital instrument for such flying. Bonanza N3794N was fitted with an older 'Sperry F3' attitude gyroscope. These two instruments had diametrically opposite styles of display of the aircraft's pitch attitude. In the 'conventional' display, the sky is at the top and the ground at the bottom. The Sperry, however, had the sky at the bottom and the ground at the top. That was fine if one had constant visual ground references and was not solely reliant on the attitude instruments. If, however, one was concentrating on instruments to hold the aircraft's attitude, the Sperry horizon display could be counter-intuitive in inexperienced hands.

From Mason City airport, the proposed flight was a straightforward single-leg course to the north-west that would take about two hours to complete. Roger checked the weather situation at 5.15 pm on the 2nd and it was within his limits. He checked again at 11.20 pm and with conditions deteriorating, he telephoned for an update at 11.55 pm. Near Mason City airport, which stood at 1,200 feet above sea level (asl), the cloud base was now 5,000 feet asl; visibility 10 miles; temperature 17 degrees F; wind 20mph, gusting to 30mph. Around 12.55 am, while taxying for departure, Roger re-checked the weather but it was only getting worse. Cloud ceiling was down to 3,000 feet asl; 6 miles visibility; light snow flurries in a gusty wind of up to 30 knots. It appeared still possible to fly the route, but only if visual sight of the ground could be maintained. The effective cloud base was down to 1,800 feet; it was dark too – the middle of the night – and he would be flying over sparsely populated country with few ground lights for help. Furthermore, the weather was also deteriorating along the route north. Critically however, 'flash' warning reports, from the US Weather Bureau in Minneapolis, of this rapidly changing *en route* situation were not relayed by Air Traffic Services to Roger Peterson during his weather enquiries. Later, Jerry Dwyer told the enquiry that when he had accompanied Roger to check the weather, no information was given to them that indicated instrument flying weather would be encountered along the route. These facts alone made it a virtual certainty that he would enter an IMC situation – in which he was both unqualified and unauthorised to fly.

Roger Peterson opted to carry on with the flight and Jerry Dwyer did not dissuade him. They pulled the Bonanza from its hangar to await the passengers. Arriving at 12.40am, they climbed aboard; Ritchie and the Bopper taking the back seats and Buddy up front with the pilot.

Using runway 17, Roger took off and made a climbing 180-degree turn onto a course for Fargo, while Jerry Dwyer watched the departure from the control tower. He watched the tail light climbing to what he estimated was about 800 feet, then was puzzled to see it descend and disappear from his view. His radio calls to the aeroplane were not answered. Disturbed by this and fearing the worst, he decided to wait until first light then fly along the same course and try to find out what had happened.

Jerry's worst fears were indeed fulfilled when, at 9.35am on the morning of 3 February, he flew over the wreckage of an aircraft, smashed to pieces, only six miles north of Mason City airport. It could only be the Bonanza and there was no sign of life. There were no witnesses to this crash and no survivors. The report details and subsequent inquest details make hard reading and will not be aired further here.

The US Civil Aeronautics Board, forerunner of the present US National Transport Safety Board, investigated the crash. It was the investigators opinion that the probable cause of the crash was the pilot's unwise decision to embark on a flight which would necessitate flying solely by instruments when he was not properly certificated or qualified to do so. Contributing factors were serious deficiencies in the weather briefing and the pilot's unfamiliarity with the instrument which determines the attitude of the aircraft. When confronted, only a few miles from his take-off airfield, with a rapid change from VMC to IMC conditions, he became disorientated and lost control of the aircraft.

It speaks volumes about the way this tour was organised, that it continued uninterrupted. New acts were wheeled out to fill in for Buddy, Ritchie and The Big Bopper and the show carried on. It was a heaven-sent break for the likes of Robert Velline (aka Bobby Vee), who headlined the Moorhead show at short notice and Fabian, Frankie Avalon and Ronnie Smith, who also had an unexpected chance to shine during the remainder of the schedule. Waylon Jennings carved out a glittering career as a Country-&-Western star before his death in 2002, while Tommy Allsup: the legendary guitar maestro, became a record producer and session musician, featuring on 6,500 recordings up to his death in 2017. After the Tour, Carl Bunch joined the US Army, then returned to the music scene for a time, before being ordained as a Southern Baptist Minister. He died in 2011.

The Buddy Holly tragedy resulted in a generation of rock-'n-roll entertainers being wiped out at a stroke. But the music did not die.

Chapter 12

Dag Hammarskjöld

Dag Hammarskjöld is widely regarded as the most dynamic and influential Secretary-General the United Nations (UN) has seen. After the resignation in November 1952 of the first Secretary-General: the Norwegian Trygve Lie, the major world powers, led by the USA and the USSR, tried to manipulate the selection of his replacement by variously vetoing candidates until they settled upon one they thought would be a push-over for all of them. That person was Dag Hammarskjöld; a professional Swedish diplomat and economist, perceived as an unremarkable character; a perennial administrator who, for a quiet life, might be prodded into doing as he was told. But, the big boys were in for a surprise. In April 1953, when all the manoeuvring and horse-trading was over, Dag Hammarskjöld was elected as the second – and youngest – Secretary-General of the United Nations for a five-year term. He rose valiantly to the task and made a significant contribution on the world stage – and turned out to be no-one's lap-dog.

Born in Jönköping, Sweden on 29 July 1905, Dag Hjalmar Agne Carl was the fourth and youngest son of a former Prime Minister of Sweden (1914-1917), Knut Hjalmar Leonard Hammarskjöld and his wife Agnes, neé Almquist. Together with his brothers Bo, Ake and Sten, they were brought up in an aristocratic lifestyle in the very grand surroundings of Uppsala Castle, which was the official residence of their father while he was the governor of the province of Uppsala.

He left secondary school in Uppsala at the age of 18 years with a string of 'A'-grade passes that ensured entry to Uppsala University to study French, philosophy and economics and was awarded an economics degree and a law degree. His worldly education was greatly supplemented since, with a father who held high office in the Swedish government, Dag had excellent opportunities to mix with the intellectual elite of the land.

Aged 25, with family connections in politics and government, it was inevitable that Dag would find his way into that sphere and he found a post, first as assistant secretary of the Swedish government Unemployment Committee, then later as its First Secretary. By 1933 he had completed a PhD in economics and from now on, this highly educated man began his climb through the ranks of the Swedish government as a dedicated and trusted civil servant.

As the first Secretary-General of the UN (UNSG) the Norwegian statesman, Trygve Lie supported the aggressive line led by the United States in the Korean War and as a result, had lost the trust of the USSR, which supported the North Koreans. His standing within the UN establishment deteriorated to such an extent that he resigned in November 1952. The names of several people were put forward unsuccessfully to the permanent members of the UN Security Council, whose job it was to select a new Secretary-General. Dag was not well-known, but he was proposed by France and finding favour with both USA and USSR, he was elected as the second United Nations Secretary-General (UNSG), with effect from 10 April 1953.

Dag Hammarskjöld never married and from the energy he directed at his new task, it might be said that the only thing he was wedded to was his job. He enjoyed the outdoors, particularly hiking among mountains and amidst the loneliness inherent in his position, he found solace in reading and poetry. Unusually, he wrote many of his own speeches as UNSG, although he is generally considered to have been an uninspiring orator.

Based in the UN Headquarters in New York, Dag Hammarskjöld set down his own markers about what he believed the UN objectives should be. In his opinion its main purpose should be to keep the peace, which it would achieve through nations having respect for international law and through the employment of the UN's supervisory capability as an internationally-recognised independent organisation. He felt that having or promoting peace would form a basis for social progress throughout the world. Over time he would find that not all of his world-contacts were in tune with his own idealism, intellect or high standards, but he remained single-minded in his stance.

It was not long after Dag had been sworn in, that the Korean War ended on 27 July 1953. In the aftermath, Dag had his first opportunity to exert his strength of character, personality and diplomacy upon a delicate political situation involving the fate of American prisoner of war (POW) aircrew who were held by the Peoples Republic of China for what it alleged were violations of its air space. It was a vindication of what Dag called his Peking Formula, in which the UNSG had acted personally as a facilitator of solutions through a policy of quiet diplomacy, that Chinese leader Chou En-Lai announced that the convicted pilots would be released on 1 August.

The United Nations was drawn into tidying up the mess of the Suez Canal crisis in 1956, but this gave Dag Hammarskjöld an opportunity to put into practise one of his 'supervisory' visions and he negotiated tirelessly with member states to create a peace-keeping force. For the first time, the UN provided an armed peace-keeping force, known as the United Nations Emergency Force #1 (UNEF 1), which comprised of 6,000 troops drawn from ten member nations, with a UN appointed General in command. The troops wore their own standard clothing but were issued with a UN-blue beret, helmet and shoulder badges. The adoption of blue headgear was Hammarskjöld's idea and its use persists to this day. This force was sent to Egypt, first to oversee the withdrawal of occupation forces and then to act as a buffer – with the consent of the two parties – between Egypt and Israel in the Canal Zone, Sinai and Gaza areas, where UN troops remained with great success and impartiality, until they finally withdrew at the request of the Egyptian government in 1967. The Canal was handed over to Egypt to run and it was re-opened to shipping in March 1957. The outcome of the Suez episode was a success for the UN and it enhanced Dag's reputation on the world stage.

While the world was focussed on Suez, Hungary entered a state of revolution. Dag received some criticism for the impotent UN response to Soviet aggression in Hungary, but it required all the parties to meet and talk about solutions and since the USSR flatly refused to engage, he was in an impossible situation. His personal diplomatic style, energy and expertise, however, were never in doubt and still seen as the best candidate for the post of Secretary-General, in 1957 he was unanimously re-elected for a second five-year term. He found the job highly stressful but said he saw it as his duty to continue. Setting out his future vision for the UN, he felt its actions should aim to be pro-active rather than re-active. As an economist, Dag pledged to promote the need for the UN to re-prioritise the distribution of financial aid so that poor countries felt the benefit. By doing so, he believed that economic development would encourage de-colonialisation and lead to more stability in the world. While an admirable ideal, the issue of de-colonisation was fraught with danger and would come back to haunt Dag in years to come.

In the words of UK Prime Minister, Harold Macmillan, delivered in a speech in Accra, Ghana on 9 January 1960: 'the wind of change is blowing right through Africa. This rapid emergence of the countries of Africa gives the continent a new importance in the world.' De-colonisation was here to stay and the UNSG would become progressively more involved.

In 1960, nationalism in the Belgian Congo had brought about – somewhat prematurely for both participants – independence from its former master, Belgium. As the Republic of the Congo, the regime of President Joseph Kasavubu and Prime Minister Patrice Lumumba found the task of self-government far from easy and factional and ethnic unrest soon reared its head. Belgium, while acquiescing to independence, was reluctant to let white-rule go completely. In this mineral-rich country, in addition to the vested interests of international business, the major eastern and western political blocs were also involved since, beside it being a major world source of copper, the region also had deposits of uranium, vital to their nuclear weapon programmes.

The southern mineral-rich province of Katanga, under its governor, Moise Tshombe, expressed complete dissatisfaction with the central government's chaos and declared itself an independent state. When Tshombe's move was supported by troops and civil servants from Belgium – which held substantial interests in Katangan mining – central governance of the Congo started to crumble. In August, neighbouring South Kasai, also a mineral-rich mining area, was next to declare independence and Belgium – with its mining interests also in mind – lent its support to this move, too. Losing these two provinces cost the Congo 40 per cent of its state revenues. The Congolese government asked the UN to provide it with military assistance and Dag Hammarskjöld took personal control of the issue.

The Congo marked the zenith of Dag Hammarskjöld's tenure as UNSG but, it was in his handling of the crisis that Dag seemed to move away from the degree of impartiality that had been his hallmark. He wanted the Congo to receive whatever aid it needed but he was adamant that it had to be sourced multi-laterally and under the scrutiny of the UN. Furthermore, he was determined to prevent Soviet influence in the governance of the Congo and its assets. To deal with the latter, he adopted a strong interventionist stance, with the insertion of a UN military force known as *Opération des Nations Unies au Congo* (ONUC), which would oversee the removal of Belgian troops, act as a peace-keeping force and in Dag's words: 'keep the Cold War out of the Congo.'[1] Dag appointed the Swedish economist Sture Linnér as the chief of UN operations in Congo (UNOC).

Despite Dag's wish to keep Katanga as an integral part of the Congo, he felt he had to resist calls for more intervention by way of sending in UN forces to take over control in the province. In June 1961, the situation changed following his appointment of the Irish diplomat Conor Cruise O'Brien as his representative in Katanga. Now, with the application of firmer diplomatic mediation, the unrest that had been fuelled by Belgium, Britain and France and a large influx of mercenaries, calmed down somewhat. In a crackdown by UN forces during late-August, about forty pro-Katangan mercenaries were rounded up and deported. Then, on 11 September, O'Brien was authorised by the UN mission to Congo to arrest senior Katangan politicians, including the self-styled President, Moise Tshombe. However, Tshombe escaped to the neighbouring British colony of Northern Rhodesia (now Zambia). On 13 September, UN forces set about taking control of Katanga by aggressive force in Operation *Morthor*, with the objective of subduing the Katangan military; clearing out European and mercenary forces and re-integrating the secessionist province with the Republic of the Congo. Using Belgian mercenary pilots, the Katangan side made good use of a Fouga CM-170 *Magister* jet trainer, one of three, ordered by Belgium; supplied by Potez France and delivered during February 1961 in an air freighter operated by Seven Seas Airlines – an American company with alleged links to the CIA.

Dag Hammarskjöld flew from New York to Leopoldville (now Kinshasa) on 12 September, arriving the next day, 13 September to discuss the implementation of a programme of aid with the central government of the Congo. However, while he was in Leopoldville, he became acutely aware of the fighting that had broken out between UN and Katangan forces. The UN had suddenly moved from peace-keeping to peace enforcement.

When his efforts to broker a ceasefire failed, on 16 September Dag contacted President Tshombe through British diplomatic channels and sought a meeting. He felt he could still extricate himself from this situation by negotiating a peaceful solution to the Katanga crisis. He felt it would be hard, even for the notoriously fickle Tshombe, to reject a personal proposal for peace from the UN Secretary-General himself. Tshombe, having fled to Northern Rhodesia, indicated he would meet Dag on 17 September in Ndola.

A Douglas DC-6B, SE-BDY, *Albertina*, of TransAir Sweden, leased by the UN for the use of the ONUC Force Commander, had flown in to Ndjili airport, Leopoldville, from Elisabethville on the morning of 17 September and was put at the disposal of the UNSG for his trip to Ndola. SE-BDY was originally delivered to ARAMCO as N709A on 17 July 1952. It was sold to TransAir on 1 August 1961, but never saw service in Sweden because, from that date, it was leased from TransAir by the United Nations and flown direct from the USA to Congo.

With a wary eye on the existence of the Katangan Fouga fighter, for security reasons a spurious destination of Luluabourg was noted on the obligatory flight-plan. The secret flight plan for the DC-6 involved taking a longer, dog-leg route, east to Lake Tanganyika then south to Ndola, thus avoiding Congolese air space for a large part of the journey, over which radio silence would also be maintained. The total air distance to be flown was 2,325km and the aeroplane was fuelled to its maximum for the flight. The three pilots on board the DC-6 were all highly qualified for both day- and night-time flying – and the captain was an expert navigation instructor, too – but none of them had flown into Ndola airport at night.

The aircraft cabin was searched, then Dag Hammarskjöld and his 'number-two,' Sture Linnér, boarded the DC-6. Sitting side by side; seat belts fastened and with the sound of the aeroplane's engines being fired-up, suddenly Dag turned to Linnér and said they really should not both be away from Leopoldville at such a fragile time. He then ordered Linnér to leave the aeroplane and stay behind. The aircraft steps were still in place and both Linnér and Dag left their seats and went down onto the tarmac. In the noise and prop-wash, the two men spoke for a couple of minutes at the foot of the stairs, then Linnér walked away to the terminal while Dag re-boarded the aeroplane.

At 15.51 GMT on 17 September, Dag Hammarskjöld and his party of fifteen (see Appendix 4), took off from Leopoldville airport. At 20.35 GMT, the DC-6, having entered airspace controlled by Salisbury (Rhodesia) Flight Information Centre, sent a radio message that the aircraft was over the southern end of Lake Tanganyika, on a southerly heading, outside Congo and Katanga territory. Given clearance to descend to 16,000 feet, this was confirmed as reached at 21.15 GMT. Salisbury handed the DC-6 over to Ndola control at 21.32 GMT. Now approaching from the south-east, the DC-6 contacted Ndola at 21.35 GMT and gave an estimated time of arrival (ETA) overhead Ndola as 21.47, with a landing ETA as 22.20 GMT. Information about airport weather, barometric pressure, surface wind direction and strength was passed by Ndola control tower to the DC-6 and duly acknowledged. At 21.57 clearance was given for the DC-6 to descend to 6,000 feet. At 22.10 GMT, the DC-6 crew reported by radio that it was 'abeam Ndola, airport lights in sight'. The aircraft was seen to pass by the airport heading roughly 280 degrees, which would suggest that it was beginning the standard approach pattern for this airport. There was no further radio contact with the DC-6 and it never arrived at its destination.

The DC-6 carrying Dag Hammarskjöld and his party crashed when the aircraft struck tree-tops at 4,357 feet above mean sea level (amsl), 9.5 miles from the runway, after the aircraft had turned onto a heading of 100 degrees.

It is normal practise to use a published standard approach pattern for a landing at Ndola – and indeed at all airports. In those days, an aircraft inbound from the east should pass overhead Ndola airport (code FLND – NLA, located in Ndola's Itawa district) on a heading of 280 degrees at 6,000 feet altitude in the direction of the Ndola Non-Directional Beacon (NDB). This is a radio-navigation transmitter located at a known fixed point on the ground; used to guide aircraft during an approach to landing. There is a visual 'compass-face' dial on the aircraft instrument panel with a needle that, when the radio-frequency of a particular NDB is selected, will point towards the beacon and indicate its position relative to the aircraft. Thirty seconds after passing the NDB, a left-hand 'procedure turn' (a reversal of course) should be executed to point the aircraft directly towards the Ndola NDB inbound on a heading of 100°. If, on the outbound track, the aircraft passed close to the NDB, this manoeuvre would involve an aircraft of the size of the DC-6 first executing a gentle right-hand turn, followed by the left-hand turn to bring it round onto a track to pass over the NDB inbound. Only upon completion of the procedure turn, should a descent begin to bring the aircraft down to the prescribed 5,000 feet minimum over the beacon, at which point the aircraft would be aligned with the runway and 2½ miles from touch down. Thereafter, a normal descent angle of about 3 degrees would bring the aircraft down to the runway threshold at 4,160 feet amsl.

Several investigations of this crash – in which there were no survivors (save for one person who died some days after rescue without gaining consciousness) – were conducted by different organisations, each of which examined facts, reports and theories – same of the latter being of the 'conspiracy' type – in an effort to establish the cause. These were conducted and reported upon in the immediate aftermath and variously up to as recently as 2019.

One report that seems to cut a sensible path through conjecture, is the comprehensive evaluation of the factual information and data gathered immediately after the crash, conducted in 2013 by Sven Hammarberg. It is significant that he is a professional air accident investigator and adopts that role for this investigation. His analysis points firmly to the cause being Controlled Flight into Terrain (CFIT) – i.e. an accident without an external stimulus. Many of the other investigations seem unwilling to acknowledge that the cause is most likely to have been an error on the part of the flight-deck crew. These other 'popular' theories: sabotage; a bomb explosion; fuel shortage; brought down by ground fire; an air-interception shoot-down; wrong QNH; use of incorrect airport approach data and faulty altimeters, while always impossible to completely rule out, are simply not supported by the crash data. One of the most popular of the conspiracy theories, for example, has been the 'shoot-down' story. This, though, is convincingly discounted by Hammarberg. It rears its head again in 2019 after much newspaper coverage of hearsay reports that a now-deceased Katangan mercenary and former RAF-trained Belgian pilot: Jan van Risseghem, shot down Hammarskjöld's DC-6. This story – and many of the other theories, were aired in yet another report to the UN Security Council, dated 12 September 2019, written by Mohamed Chande Othman, a Tanzanian Chief Justice and head of a UN panel of experts charged with the assessment and examination of new information relating to the death of Dag Hammarskjöld. He, too, ruled nothing out; was clearly reluctant to accept a CFIT conclusion and recommended even more investigation. Hammarberg, though, summed it up as follows.

'A number of facts are pointing at (or are totally consistent with) CFIT, at the accident site and the wreckage as well as in surrounding details. The crash site is located roughly under a normal procedure turn to the airport. The swathe cut in the trees shows that the aircraft hit the trees in a very shallow descent, approximately in accordance with a normal descent towards the runway. The first impacts were made by the aircraft's propellers. There were no signs of any impact, or aircraft parts, before this point. The landing gear was down and locked, and the flaps in approach position. All engines were working at impact, approximately at approach power. The altimeters were correctly set, and no pre-impact damage was indicated or found. The [aircraft's] instrumentation standard at the time was not sufficient for safe flight in poor visibility.

My conclusion is that there is no need at all for external disturbances or hostile acts to make an accident look exactly like what we see in the SE-BDY case. Thus, 'the attack' is not needed. The 'other aircraft' is not needed. No bombs, no world-wide conspiracies are necessary to bring down an aircraft. Actually, all the factual findings point toward another, much simpler, conclusion.

The visual conditions were good concerning horizontal visibility, but with high risk of illusions known [in these modern times] to cause lack of awareness of height above ground. The landing chart didn't support full safety in that it didn't show the high ground level west of the airfield and it could be interpreted as to permit descent to 5,000 feet too early. The crew was, in the circumstances, to fall into an illusion trap of [thinking they were] at safe altitude, caused by the 'black hole' conditions.

Fatigue after a long flight may very well have contributed to the accident, especially in combination with the relief of having reached the destination. It is possible that a higher overall safety level would have revealed the risks associated with this particular flight. It is clear, and ironic, that flight safety in this case was degraded because of security precautions. The classification as a CFIT does not mean that the crew was incompetent or ignorant. Instead, had the world's aviation community earlier promoted in-depth analysis of causes to accidents and had reports not been concluded with just: 'pilot error,' the risks connected to flying over dark areas could have been known to the crew, thereby preventing the accident.'[2]

If, after sixty years, no-one has still been able to point the finger with any certainty at a politically – or commercially-motivated assassination, then the 'truth' probably does lie in the simplest explanation.

Chapter 13

June Thorburn

While making a controlled descent into London Heathrow airport on a cold, misty night in November 1967, Iberian Airlines Flight IB-062 inbound from Malaga, Spain, plunged into Blackdown Hill in West Sussex, killing all on board. One of the passengers on that tragic night was June Thorburn, a popular, internationally-known actress, noted for her portrayal of quintessential 'English Rose' characters on television and in British and Hollywood movies. Starring in more than twenty dramatic, musical and comedy feature films, she also featured in over sixty British television shows and series-episodes and was a regular on the London West End theatre stage.

Born on 8 June 1931 in Karachi (now in Pakistan), Patricia June Thubron Smith was the eldest of three children; daughter of a British officer in the Indian Army, Colonel R.G. Smith and his wife Emmeline (neé Thubron). Patricia (June) had a sister, Diana and a brother, Keith. June eventually took her professional stage name from her own second Christian name and a version of her mother's maiden name of Thubron. In 1951, her first theatrical agent suggested that no-one in show-biz would be able to pronounce or spell 'Thubron,' so they settled upon the simpler version: 'Thorburn' and thus, she became 'June Thorburn' to her public.

When the British Indian Army was disbanded in 1947 in the wake of Indian Independence, Colonel Smith and his family returned to England, settling into an English way of life in Fleet, Hampshire. Having left school, June, who had by then matured into an attractive young woman, decided to take up acting as a career. There is some evidence to suggest that, while living in Fleet, June became a member of the Royal Army Medical Corps Dramatic Society (RAMC DS), which used the Globe Theatre in Queen Elizabeth Barracks, Crookham and the Aldershot Repertory Company which played the Theatre Royal in Aldershot – both venues being close to her home in Fleet. June played other repertory theatres in Eastbourne, Brighton and Southampton but, when she pursued her acting ambitions in London, to help pay the rent she took a variety of casual jobs. Spotted by a *Daily Mirror* photographer; her picture was printed in the newspaper on 22 May 1951, with the caption:

'June Thubron (19) wanders through Covent Garden Market, London, buying flowers for her basket. June, is one of the Victorian-style flower girls at the Battersea Park Festival Gardens.'

It was while selling programmes at Battersea Fun Fair that she met Aldon Richard Bryse-Harvey and married him that same year. Little is known about Bryse-Harvey, except to say that he changed his names by deed poll from Albert Harvey Walkey, to Aldon Richard Bryse-Harvey on 18 December 1951. Quite why he did this, is not known.

Following her film debut in *The Children of Camp Fortune*, June made her West End stage debut in the London production of *Red Letter Day*, that ran at the Garrick Theatre in Charing Cross Road from 21 February to 24 May 1952. She played the relatively minor role of 'Alice' amongst an illustrious cast that included such heavyweights as Donald Sinden, Hugh Williams, Terence Longdon and Fay Compton. June is listed in the cast as 'June Thorburn,' which suggests she had settled on her stage name sometime between May 1951 and February 1952. It was in this play that June was spotted by a famous freelance film casting director named Maude Spector.

June's big break into feature films came later that year when Maude Spector secured a small supporting role for her in the British film *The Pickwick Papers*. Premiered on 14 November 1952, this film, made by Renown Picture Corporation at Nettlefold Studios in Walton-on-Thames, brought together an outstandingly talented cast, headed by James Hayter. June – who at the start of filming was six-months pregnant but showing no outward sign of her condition – played 'Arabella Allen'. She made a good job of the role and among the many people in the film profession who asked: 'who is that pretty girl?' was Sir Michael Balcon, who signed June up for one of the few supporting female parts in *The Cruel Sea*, a dramatic, raw-edged war film set aboard Royal Navy convoy-escort vessels during the Battle of the Atlantic. In this film June portrays 'Doris Ferraby,' the pretty, young new wife of 'Sub-Lieutenant Gordon Ferraby,' played by John Stratton.

During her career, June enjoyed a long association with television and at this early stage she landed a couple of parts in two episodes of a long-running drama series hosted by the Hollywood legend Douglas Fairbanks Jr. His series of dramas, mysteries and farces was aired on BBC TV under the title of *Rheingold Theatre* which ran from 1953 to 1957. The birth of her first child came on 1 May 1953: a daughter named Heather-Louise June (Bryse-Harvey) and it is believed Fairbanks acted as godfather to the child.

The year 1954 would prove to be a busy one for June, with an eclectic mix of film genres. First up was the British movie *Fast and Loose*, an adaption of a Ben Travers farce, made by Group Film Productions Ltd and released in February. July that year saw the release of the British movie: *Delayed Action*, a *film noir* drama from the Kenilworth Film studio, in which June appeared in her first leading lady part. In August 1954 she was back on British television in an episode of the massively popular *Sunday Night Theatre*, a 90-minute live play series produced by BBC TV that ran between 1950 and 1959. June played the supporting part of 'Nellie Sellenger,' daughter of 'Lady Sellenger,' played by that *grande dame* of stage and screen, Fabia Drake. Next came a role in another comedy/farce called *Orders are Orders*, by Group 3 Studio. Released in October 1954, it starred Sid James, Tony Hancock, Peter Sellers and Brian Reece, with June Thorburn appearing in a minor role as 'Veronica Bellamy'.

It was a glittering occasion on 31 October 1955 when June was one of dozens of international stars assembled in the Odeon cinema, Leicester Square, London to meet HM The Queen for the Royal Film Command Performance of *To Catch a Thief*, a Paramount Picture directed by Alfred Hitchcock and starring Cary Grant and Grace Kelly. Standing next to Peter Ustinov in the line of stars, June looked radiant and made a deep curtsey as she shook hands with The Queen.

Sadly though, it was around this time that, as her acting career blossomed, her marriage began to founder. During 1955, June and Aldon Bryse-Harvey were divorced and she moved back into her parents' home in Crookham, Hampshire with her daughter Heather-Louise and a children's nanny.

During late 1956 June secured a leading role in a Rank Films nautical comedy called: *True as a Turtle*. *The Times* reviewer thought this film tried to do for sailing what *Genevieve* did for veteran cars – but sadly with less success! Released in February 1957, June plays 'Jane Hudson', the new bride of 'Tony Hudson', played by John Gregson. They suffer many misadventures and discomforts but love, of course, triumphs in the end.

During the rest of 1957 and into 1958, June worked a lot in television. She appeared in shows such as *The Shadow Squad*, an ITV drama series by Grenada Television that ran twice a week between 1957 and 1959. On 30 January 1958, she appeared in a 90-minute drama film for BBC TV called *Shut Out The Night*, with the actor Paul Eddington. June also appeared in the hugely popular ITV *Play of the Week*, a stalwart of 1950s television that aired between 1955 and 1964.

For the remainder of that year, June worked on her first Hollywood/UK co-production. *Tom Thumb* was a fantasy/musical film by MGM & Galaxy Productions and was to be made partly at MGM's studios in Borehamwood, London and partly in Hollywood. A further innovation was that it would use a mix of live and animation filming. The American actor, dancer and acrobat Russ Tamblyn had the title role, but almost all of the supporting cast was drawn from well-known British actors. June Thorburn was cast in a major supporting role as the mystical 'Forest Queen' and seemed to be quite at home in that romantic part. Released in December 1958, *Tom Thumb* enjoyed good reviews; made a profit at the box office and became the eighth most popular film in Britain during 1959. *Variety* magazine considered the film 'a top-drawer [production] that equals some of the Disney classics;' while *Time* magazine commented that 'it was unusually fresh and appealing.'

For June, these parts continued to roll in and her reputation as an international movie star grew steadily. It was off to sunny Spain for June in April 1959, when she landed a leading lady part in the film, *The Three Worlds of Gulliver*, a Columbia Studios production. Between April and June, exterior filming was done on and around the beach near S'Agaro, on the Costa Brava in Spain, then the cast and crew returned to Shepperton Studios in London to film the interior sequences. This film broke new technical ground by utilising 'stop-motion animation,' publicised under the grand name of 'Super-Dynamation.' It was used, for example, to 'shrink' some of the performers to 'Lilliption' size and the process was overseen by that master of special effects in those pre-computer-graphics days: Ray Harryhausen.

For June, the year 1960 turned into a conveyor-belt of TV films, plays and series. Her shows included: the TV film *Meeting at Night*; two episodes in ITV's *Television Playhouse* and one episode in the ITV series *International Detective*. She appeared in three episodes of *Tales of the Vikings*, an American TV series broadcast in the USA and was produced by Kirk Douglas's own company: Brynaprod.

One of June's biggest TV shows was the British drama series called: *The Four Just Men* made by Sapphire Films at Walton-on-Thames studios. Its stars were from the Hollywood top drawer: Richard Conte (US), Dan Dailey (US), Jack Hawkins (UK) and Vittorio De Sica (Italy). June Thorburn played 'Vicky,' a law student assistant to Richard Conte.

There was further recognition of her professional standing when June was invited to be a 'castaway' on *Desert Island Discs*, the long-running BBC radio programme created in 1942 by Roy Plomley. 'Castaways' are gently interviewed about their life and work and choose eight pieces of music to take with them to the 'island.' They are also allowed to take a luxury item and June chose a mink stole! Her programme was broadcast on 2 February 1959 on the BBC Home Service.

It was around mid-1959 that June began dating a new man in her life, a Norwegian shipping businessman: Morten Smith-Petersen, a director of the Bergen Shipping Line. On or about 23 March 1960, June and Morten married in a quiet ceremony at Hampstead Registry Office and went on a honeymoon cruise to North Africa. So, at the age of 28-years, June now became Mrs Smith-Petersen and also gained a step-son named Simon, from her husband's previous marriage.

It is believed that having made several films with Pete Murray, when he moved into the music and 'disc-jockey' industry, he arranged for June to be invited onto Juke Box Jury, on which he appeared quite regularly. Presented on BBC TV on Saturday evenings, by its host David Jacobs, this popular music panel-show drew 12-million viewers at its peak. June appeared in ten shows between 1961 and 1965, sitting alongside such household names as: Frankie Vaughan, Pete Murray, Tony Bennett,

Ted Heath, Buddy Greco, Bobby Vee, Terry-Thomas, Arthur Askey, Sabrina and Jackie Collins, to name but a few.

June continued her steady output of TV films and series and was always in demand at social and publicity-related events. She was equally at home, for example, doing charity events at Battersea Festival Gardens (1963); attending Variety Club charity days at York and Sandown Park races (1962) – with a bevy of fellow British film stars; attending the film premiere of *Don't Bother to Knock* in the brand-new ABC Cinema in Sheffield in May 1961 – or being seen on the French Riviera at the 1962 Cannes Film Festival. Arising out of her work with the Irish Ardmore Studio, one unexpected pleasure came in the form of an invitation for June and her husband to attend a garden party in Dublin, given by Irish President Eamonn de Valera to honour a visit to the republic by US President John F. Kennedy in late-June 1963.

On 10 May 1964, June gave birth to her second child, a daughter: Inger-Sheleen Christabel (Smith-Petersen). Once again there is a period when she had to play the role of 'mother' rather than go off to work in films, but she was back doing TV work in 1965. June also found time to take to the stage again; returning in 1966 to the West End in George Bernard Shaw's witty comedy: *Man and Superman* at the Garrick Theatre. Playing 'Ann Barker' on 7 October 1966 in an episode of *Blackmail*, an ITV drama series that ran from 1966 to 1967, appears to be June's last public role.

It was on 4 November 1967 that the flight which ended her life, took off from Malaga airport. No clear explanation emerges as to why she had gone to Spain in the first place, although several newspapers reported that: 'most passengers were late holidaymakers and a few businessmen.'

Iberia Airlines flight, IB-062, was operated with a French-built Sud-Aviation SE-210 *Caravelle* 10R twin-jet airliner, registration EC-BDD and named *Jesus Guridi* after a famous Spanish/Basque music composer. It departed Malaga airport at 19.30 hours GMT on Saturday, 4 November 1967, bound for London Heathrow airport, where it was due to land at 22.10 hours GMT. With a capacity for up to eighty passengers, this scheduled flight carried thirty passengers and a Spanish crew of seven (see Appendix 5), under the command of 37 years-old Captain Hernando Maura Pieres.

The flight was proceeding quite normally as the *Caravelle* approached latitude 50 degrees North. Cruising at 31,000 feet it was cleared by Air Traffic Control (ATC) to descend to an altitude of 21,000 feet and route via Ibsley (New Forest) and Dunsfold radio navigation beacons. It was then instructed to descend further, to 11,000 feet and directed to turn right onto a course of 060 degrees towards Dunsfold beacon (south of Guildford). Now passing abeam of Fawley (Southampton), the pilot was instructed to descend to 6,000 feet altitude. When acknowledging that instruction, the crew reported the aircraft was descending through 14,500 feet altitude. Four minutes later, the aircraft was now cleared to proceed on its own navigation, directly to the Epsom beacon on a course of 050 degrees. This instruction was acknowledged but, just 1½-minutes later, at 22.02 hours, the aircraft hit the ground on Blackdown Hill, near the village of Fernhurst, in Sussex. The point of impact was at a height of 700 feet above mean sea level (amsl) and 200 feet below the summit of the hill, the highest point in Sussex.

It was a drizzly, slightly misty night, with low cloud and there were a number of Bonfire Night and firework events across the country. No-one in the vicinity actually witnessed the impact, but several people telephoned the police and fire brigade to report that they had heard a loud explosion in the hilly part of the Haslemere/Fernhurst area. Air Traffic Control (ATC) also reported their loss of radar contact to Sussex Police HQ, indicating their plot had disappeared in that area, too. The Hascombe village policeman, whose beat covered that area, was off-duty, but he was asked to check-out the report. Knowing the area well, he headed-off on his motorcycle to Blackdown Hill, as a vantage point from which he might be able to see what was going on. All was silent, but there was an acrid smell in the air and he saw small fires dotted around in the darkness down the hill. With no radio and just a hand torch to light his way, he set off into the fields below where he soon stumbled over dead sheep. Further on he found pieces of metal wreckage lying all around and moving into the trees and undergrowth, he found horrific evidence of human wreckage, too. He was joined by another policeman who, guided by his colleague's torch and shouts and a strong paraffin-like smell, passed the burning bodies of sheep and into the swathe of wreckage. Several police patrol cars were dispatched to the general area and guided by the fires, quickly reached the location. Proceeding in darkness on foot, helped on by shouts from their colleagues, a scene of terrible devastation unfolded for 800 yards in front of these police crews. Clambering inside and among wrecked parts of what was clearly a crashed aeroplane, they shouted and searched for survivors – but they searched in vain. As daylight dawned, road blocks were set up to prevent public access as the macabre extent of the devastation became apparent. Police, fire brigades and ambulance personnel could be seen picking their way through the site, but they were searching now for human remains, luggage and personal belongings. A Board of Trade accident investigation team, led by Group Captain J.B. Veal, was also beginning the painstaking task of trying to establish what caused this tragic event. One of the key tasks was to find the Bendix flight data recorder, sometimes known as the 'black box,' which might give the team vital clues.

Perversely, it was lucky the airliner had hit the hill. Here there were woods but almost no houses, whereas surrounding the hill were several housing estates. First, the aeroplane broke off the tops of trees in the grounds of Blackdown House – missing the house, with its elderly occupants, by a few yards – then it skidded for a quarter of a mile across a field full of sheep. Sixty-five animals were killed and twenty-three more were so badly injured that they had to be put down later. At a forward speed of what must have been close on 300mph, shedding wreckage, the airliner now demolished a garage belonging to Upper Blackdown House and its starboard wing struck and damaged the roof and a wall of this house. On it swept, before finally ploughing into dense trees and undergrowth of the woods where the wings and fuselage to the rear of the cockpit were shredded to pieces. The cockpit area came to rest with its nose embedded in the soft soil; its three lifeless occupants found strapped into their seats.

The Board of Trade accident investigation report, CAP 343, stated that at 21.42 the crew called ATC over Alderney (Channel Islands) at 31,000 feet. They were instructed to descend to 21,000 feet, which appears to have taken about 6 minutes. From 21,000 feet altitude the descent continued steadily until the aircraft struck the ground – at roughly 700 feet amsl – 13½-minutes later. Although some segments of the descent were made at slightly different rates, this represents an overall average rate of descent of about 1,500 feet per minute.

In contrast to modern times, it might be asked why the air traffic controller did not advise the pilot that his altitude was becoming dangerous? At this point in time, the radar display seen by the air traffic controller did not show the altitude at which an aircraft was flying. This integrated facility did not become available in the UK until around 1969 and thus monitoring of altitude was solely in the hands of the flight-deck crew.

Much space was given in the report to the altimeter and its possible mis-reading by the pilots. The altimeters installed in this aircraft were of a 'three-pointer' type – a not uncommon model at that time – the hands of which would be moving constantly under these conditions. This model of altimeter has an additional aid to determining the status of a descent. It comes in the form of a small aperture in its dial, which is gradually filled – in the manner of a drooping 'eyelid' – by a 'cross-hatching' indicating that the aeroplane has passed below 26,666 feet. At an altitude of 10,000 feet, this cross-hatching completely fills the aperture (the eyelid is closed) and it remains filled while the aeroplane remains below 10,000 feet. Investigators considered that this cross-hatching signal would have been visible to the pilots for at least 9½-minutes.

After considering that the aeroplane appeared to experience no aerodynamic, power or mechanical faults and was on course and under control, it was the conclusion of the investigators that at some unidentified stage, the pilots mis-read the altimeter in a way that caused the aeroplane to fly below the flight levels assigned by air traffic control. It was speculated that the composite way in which the three altimeter hands needed to be read to obtain an altitude reading, led to an error of 10,000 feet too low, possibly due (but not proven) to a mis-reading of the position of the pointer indicating the 10,000 units. It appears too that, at a crucial point, the significance of the cross-hatch indicator was overlooked (or possibly malfunctioned?). Due to the darkness and low cloud, there was also no opportunity to check their height by visual reference to external lights or ground features.

The investigators could find no explanation for the actions of the pilots and concluded that they bore responsibility for the crash. One positive outcome of the crash was that this model of altimeter was subsequently replaced by easier-to-read models in aeroplanes of this and other airlines.

Nineteen of the victims were buried in a communal grave at Brookwood Cemetery, Surrey, including one of the Spanish crew and two Spanish passengers. This plot is marked by a granite memorial plinth, surrounded by slate headstones; one for each person. Twelve passengers and six of the crew are buried elsewhere, including June Thorburn (Mrs June Smith-Petersen) who was cremated privately, following a funeral service held at Christ Church, Hampstead Square, London NW3, on 14 November 1967.

Chapter 14

Yuri Gagarin

27 March 1968

In what must rank as one of the most historic broadcasts ever made, at 10.02 am, Moscow time on 12 April 1961, the Soviet news-agency TASS announced: 'On April 12, 1961, in the Soviet Union, there was launched into orbit around the earth, the first satellite-ship with a person on board.'

106 minutes in space was all it took to turn Russian cosmonaut Yuri Alekseyevich Gagarin into the most famous man in the world, on the day the Soviet Union launched a rocket carrying the first human being into a true space orbit that encircled the earth. This epoch-defining voyage lasted for just a single circuit yet, when Gagarin landed, the world would never be the same again. The cold war was at its height; the space race between the USSR and the USA was hotting-up, and in the race to put the first man into space – to the chagrin of the astronauts, politicians and people of the USA – the USSR had won the first round convincingly.

Yuri Gagarin, a man from a humble peasant background, through his own efforts became so famous in every corner of that globe round which he had flown, that his name and unique achievement remains in the forefront of public consciousness today, almost sixty-five years on. Born on 9 March 1934, in Klushino, a rural village in the Smolensk region, 100 miles west of Moscow, he was the third of four children of parents who worked on one of Joseph Stalin's infamous collective farms. His father, Aleksei, pressed by the regime into service as a farmer, was by trade a skilled carpenter who could turn his hands to almost any practical task. His mother, Anna, held the home together; ensuring her flock were clean, tidy and well-disciplined, while passing on her enthusiasm for reading and learning to the children.

Their tiny, but relatively comfortable world imploded in June 1941 when the Germans invaded the USSR. Klushino – or what little was left of it after bombing and shelling – was liberated by the Soviet army in the spring of 1943. With the village in ruins and a large part of its population having left, Aleksei found work in the neighbouring town of Gzhatsk, ten miles away. His carpentry and handyman skills were now in great demand as the Russians tried to rebuild their lives after the devastation of occupation. Perhaps his mind would have rested much easier if he only knew that, in 1969, Gzhatsk would be re-named Gagarin, in honour of one of his own sons.

In 1949, aged 15, Yuri reached the last year of high school and it was time for him to consider his future career path. His father hoped that Yuri would follow his brother Valentin into what had become a modest, yet flourishing family construction business. Yuri had other ideas. In a post-war Russia striving for economic growth, the lure of the big city offered a chance for him to enter a brave new world of technology, industry and social mobility that he would never taste among the peasantry of rural Gzhatsk. Unable to find a vacancy at the elite trade schools in Moscow, he had to be content with a live-in place, on a two-year course for apprentice foundrymen, at Vocational School No. 10 in the less salubrious industrial suburb city of Lyubertsy.

Finishing his trade school course in 1951 with an 'A' grade pass and glowing reports, coming up to age 18, Yuri thirsted for yet more study – punching a time clock in a factory was not how he saw his future. That summer, he enrolled on a 4-year course at an industrial technical college in the city of Saratov, one of several such schools where the upper management echelons of state industries were trained. Among the extra-mural activities offered to students was the opportunity to learn to fly at the expense of the state. Saratov Aero Club was run by a para-military sports organisation: the Volunteer Society for Co-operation with the Army, Aviation and Navy; usually known by the acronym: DOSAAF. Many of the instructors were ex-regular aircrew, with some having flown in combat during the Second World War. Its basic aim was 'the patriotic upbringing of the population and preparation of it to the defence of the Motherland.' Yuri enrolled in September 1954 and what with college studies in the mornings, organising sports in the afternoons and then attending flying training and ground school (60 hours of study) in the evenings and weekends, he was now a very busy chap, but he had at least taken his first steps towards becoming a military pilot.

By July 1955, having begun on bi-planes, he progressed to going solo in a more powerful Yakolev Yak-18 monoplane at the aero club and with over eighty solo flights in his log-book, he received his pilot licence. His service with DOSAAF exempted him from being whisked off for compulsory military national service, but now he had to make his mind up about what he wanted to do with his life. Deciding that an industrial career was not for him, he directed his efforts towards becoming an officer and a military pilot in the Soviet Air Force (SovAF). His military service began in October 1955 with a 2-year course for officer and pilot cadets at the Chkalov Military Academy, an airbase located 6 miles SW of Orenberg, a large city 900 miles SE of Moscow, close to the border with Kazakhstan.

Apparently, having begun with the familiar Yak-18, soon after moving on to jets, he had difficulty with landings and after making several rough landings, he was in danger of being kicked off the course. It seems Yuri's small stature – he was only 5 feet 2 inches in height – meant he had difficulty seeing forward out of the cockpit of a MiG-15 trainer – rather fundamental during a landing! The solution came in the form of a couple of extra seat-cushions. His landings duly improved and together with a sympathetic instructor report, he completed his conversion and went solo on jets. Yuri Gagarin's height may have been a problem here, but it would actually be beneficial in making his name later.

While at Chkalov Academy, Yuri got married, just after he graduated from flight school with his 'wings.' His bride Valentina Goryacheva was a medical student in the city, where they were married on 27 October 1957. On 7 November, Yuri was commissioned as a junior lieutenant in the Soviet Air Force and received a delightful posting to 769th Fighter Aviation Regiment at Luostari air base, on the Kola peninsula inside the Arctic Circle near Murmansk and 1,400 miles from his new bride – who remained in Orenberg for the next nine months in order to complete her own studies. Detail about his military flying remains vague but, bearing in mind this was well into the Cold War period, it seems likely Yuri would have spent most of his time patrolling the Arctic border with Norway in a MiG-15 and watching over Soviet naval shipping using the port of Murmansk. Following her own graduation in July 1958, Valentina joined Yuri in his grim northern outpost and despite the inhospitable locality and less than salubrious accommodation, their first child, daughter Yelena, was born on 17 April 1959.

It was in August 1959, when Yuri was approaching the end of his first posting, that he responded to a call for volunteers to take part in testing 'a new super-vehicle'. This was not specified in any detail but it was made clear that applicants should be fighter pilots between the ages of 25 and 30; not more than 170 centimetres (1.7 metres or 5 feet 7 inches) in height and not more than 70 kilograms (154lbs or 11 stones) in weight. As a result of this low-key advertisement, 3,500 applicants, including the 5 feet 2 inch-Yuri Gagarin, came forward. It is interesting to find that Yuri – who was typical of the potential candidates – had a mere 260 flying hours to his name at this point and was an ordinary service jet pilot with no specialist experience.

The Soviet quest to find men to send into space began with small medical teams touring SovAF bases, examining the initial applicants with the aim of reducing the field to about 350 potential candidates. These were then ordered to attend the Burdenko Military Hospital in Moscow for detailed physical, psychological and academic examinations, which grew ever more rigorous and indeed harsh, in nature. By February 1960, when the cohort was reduced to twenty, he was still in contention. Of these twenty potential cosmonauts: twelve would make it into space, four would be injured in ground accidents, three would be dismissed from the space programme for indiscipline and one would die in a training accident. Now they were posted for an indefinite period from their service units to the Central Scientific Research Aviation Hospital at Frunze/Khodynka aerodrome in the suburbs of north Moscow, to begin the cosmonaut-specific, rigorous evaluation and training regime in March 1960. After the operation moved to Chkalovsky aerodrome, about 15 miles NE of

Moscow during the summer of 1960, the technical aspects of cosmonaut training and mission development relocated to a new top-secret facility, known as the Cosmonauts Training Centre, being built at top speed in adjacent secluded woodland. Growing into the small town of Zvyozdny Gorodok, this restricted military enclave was created specifically to support the long-term Soviet space programme and became known as Star City.

Training to be a cosmonaut assumed that these 'jet-jockeys' were already familiar with ejection seats. However, they were now required to learn to use parachutes to a high degree of proficiency. Far from being directed at the more usual aspect of personal safety, this skill was in fact a vital component in the whole schedule of any of the initial Soviet manned space flights. Cosmonauts would be returned safely to Mother Earth at the end of an orbital flight by being ejected from a capsule and parachuting to the ground. Yuri began his compulsory parachute training, at an airfield not far from Saratov, in April 1960 and it was not for the faint-hearted, since in order to qualify, he and his compatriots would each make in the region of fifty free-fall descents from altitudes up to 25,000 feet, over land and water and sometimes in darkness. Weightless-ness training involved trips inside converted airliners doing parabolic flights that induced about thirty-seconds of floating weightless inside the empty cabin. Nick-named the 'vomit-rides,' Yuri did seventy-five of these trips in five days during May 1960. He also underwent high-'g'-inducing rides while strapped to a seat on the end of a rotating centrifuge gantry; extremely punishing and severely testing of the cosmonauts' physical fitness. At the other end of these extreme tests was a ten-day spell in a pressurised isolation chamber in which all form of sound and communication – except for random ear-splitting alarms – was eliminated to such a degree that some candidates cracked under the mental strain.

In January 1961, having survived all this, Yuri was selected as one of the final elite group of just six men, from whom the first cosmonauts would be drawn. Made clear what they were in for, they were asked if they wanted to return to the Air Force – or go into space! All chose space and the elite group became known as the Vanguard Six. After yet more examinations, written, oral and physical and personal presentations to the selection committee, 27-years-old Yuri Gagarin emerged as the man selected to be the first Russian in space. The final decision, made on 8 April, was greatly influenced by Lieutenant General Nikolai Kamanin, head of the *Vostok* space programme and its charismatic chief rocket designer, Sergei Korolev.

Korolev's latest beauty was the R-7 rocket, a true ICBM, standing 125 feet (38 metres) tall, weighing in at 287 tons (260 tonnes). The dimensions of this rocket were dictated by Soviet nuclear weapon characteristics and it was primarily designed to haul a warhead weighing 10,000lbs (4,500 kilograms), sub-orbitally as far as the continent of North America. It was the large dimensions of the relatively crude design of Soviet nuclear warheads, linked to the required long-range, that resulted in the correspondingly huge dimensions of the R-7 family. Korolev proposed substituting the warhead for either an un-manned spy satellite or – the idea that Premier Kruschchev liked most of all – a man-carrying capsule (*Vostok*). The USA had to develop a whole new breed of launcher – and that took time. In these simple factors lies the key to why the USSR was able to beat the USA in the initial stages of the space race.

The scene now moves to the top-secret cosmodrome at Tyuratam in the vast plains of Kazakhstan, launch site for the Soviet space programme and an arsenal for nuclear-tipped ICBMs. On 16/17 March 1961, the Vanguard Six were flown from Moscow to Tyuratam where they would see the mighty R-7 for the first time. It was a difficult parting for Yuri from Valentina because their second child, daughter Galina, was born on 7 March.

On 5 April the Vanguard Six packed their bags once again; bid their families a fond farewell and together with the mission medical and technical teams, were flown in three Ilyushin IL-14 airliners – in small groups as a safety precaution – back to Tyuratam. Comrade Korolev was about to gamble with a man's life.

Events now moved quickly, as it was thought that the Americans were getting perilously close to launching their own astronaut. On 10 April, it was confirmed that Yuri Gagarin would be the first to go up on 12 April. Before dawn on that fateful morning, Gagarin and his back-up Gherman Titov were woken; ate breakfast; were medically examined, then each was assisted into his complex space-suit with its hi-viz-orange outer layer. From the top of the huge gantry that supported the R-7, Yuri wriggled onto the contoured ejector-seat and was strapped inside *Sharik*, the spherical re-entry capsule, 7 feet (2 metres) in diameter. *Vostok* had two con-joined sections; *Sharik* and the unmanned service module that contained life-support systems and the braking rocket motor.

At 09.07 Moscow time; the thunderous roar of four boosters and the main engine ringing in his ears, with a shout of *'Poyekhali!'* (*'Let's go!'*) from Yuri to mission control, the R-7 rose majestically into a clear blue sky. Two minutes later, the four boosters separated and the first stage of the main engine continued to accelerate. Inside *Vostok*, g-forces were bearing down on Yuri's body and his heart-rate rose from 64 beats-per-minute to 150. At launch +3 minutes, above Siberia, the fairing over the *Vostok* payload at the tip of the rocket, peeled away according to plan. Strapped immobile in his seat, this now gave Yuri a limited view of the passing world through a special viewing-porthole called *Vzor* (*Look*) near his feet. In addition to 'sight-seeing,' *Vzor* had a more serious purpose. If Yuri had to take over manual control, that little porthole would be his only source of visual references for controlling the re-entry attitude.

At 09.20, mission control confirmed that *Vostok* was in an elliptical orbit (apogee x perigee) of 203 miles x 118 miles (326km x 190km), travelling eastwards towards the Pacific.

At 10.02, Moscow broadcast to the world that a citizen of the USSR had become the first man in space and that man was (newly-promoted) Major Yuri Gagarin.

Re-entry began at 10.25 when the automatic sequence controller fired a braking rocket in a 40-second burn.

10.42; the ejection warning lit up and at 23,000 feet, the hatch blew off and Yuri blasted out of *Sharik*. At 13,000 feet he separated from his seat and his parachute opened. Due to a malfunction, his reserve parachute also opened and his survival pack dangling below, broke away – but, hell, all he had to do was keep calm and he was home and dry.

Eleven minutes later, at 10.53 Moscow time, 106 minutes after launch, Yuri Gagarin, the first man in space, touched earth again – in Mother Russia – alive and kicking. Standing in a field in the middle of nowhere, in his hi-vis orange suit, wearing a huge white helmet and surrounded by a billowing orange parachute, he must have been an awesome sight for the only two people to see him touch down – an old woman and a little girl. He had landed between the villages of Smelovka and Uzmorie, 25 miles (40km) south of Saratov. The charred, but intact, capsule hit the ground at 10.48, just over 1 mile (2km) from him.

Needless to say, the next hours and days were filled with the relief and euphoria of everyone involved with the project. There was patriotic hysteria and admiration throughout the land and indeed, far beyond its shores. There, in the centre of all this, exuding genuine warmth of character and handling public adoration as if born to it, was the ever-calm, smiling, confident, immaculate Yuri Gagarin.

On 14 April, Yuri met President Nikita Krushchev at a huge ceremony in Red Square, Moscow that was followed by an almost endless round of public and private celebrations. He was made a Hero of the Soviet Union and received many other Soviet and foreign awards. As a true patriot, there was never any doubt about his communist credentials and by June 1961, he began touring the world over a period of several years, as an ambassador for the USSR, his mere presence being sufficient to project an image of the superiority of communist ideology and technology.

In the years that followed, Yuri embarked on other tours across the world and on speaking engagements all over the Soviet Union. He accepted a political post in 1962, taking up a seat in parliament as a deputy of the Supreme Soviet of the USSR. Although banned by his government from flying aeroplanes himself, he was promoted to full Colonel in the Air Force in 1963. There was a point in his pressured existence when Yuri's halo slipped. Rumours were spreading about him drinking heavily and getting into embarrassing scrapes with women – one of the latter incidents is said to have resulted in a fall from a balcony that left him with a scar on his forehead that is visible in several later photographs! However, in December 1963, he seems to have emerged from these distractions and moved back into bosom of the Soviet space programme as deputy training director at the cosmonaut training facility in Star City, Moscow.

During 1964, he commanded the cosmonaut contingent and used this position to start his own training for more space flights. He got to grips with that punishing regime and was nominated as back-up for Vladimir Komorov on the first *Soyuz* space mission, scheduled for April 1967. When Komorov was killed during his parachute descent from the *Soyuz* on 24 April 1967, Yuri was banned from making any space flights; he was too valuable an asset for the state to lose. Continuing as a cosmonaut instructor, however, made him conscious of needing the respect of these jet pilots, so he pressed his superiors to at least allow him to fly aeroplanes again and re-qualify on jets himself. In February 1968, he was authorised to begin refresher flying training and was raring to go.

By 22 March he had made eighteen dual-control flights with an instructor, totalling seven hours flying time. These were flown from Chkalovsky airfield, Moscow in a two-seat MiG-15 UTI trainer version of the MiG-15 single-seat fighter jet. His past experience meant that he got back into the swing of things pretty quickly and he just needed to be checked out a couple of times by the senior instructor before he would be sent off solo. These trips were scheduled for 27 March, with Colonel Vladimir Seryogin.

At 10.18 on 27 March 1968, Gagarin and Seryogin took off from Chkalovsky airfield in MiG-15 UTI, serial number 612739. Climbing through the cloud base of 2,000 feet towards about 15,000 feet, the plan was for Yuri to fly the aircraft and perform a series of standard, non-aerobatic manoeuvres to satisfy Seryogin of his competence to fly the MiG; the sortie only due to last about half-an-hour. The test complete; things went wrong as the MiG descended through the cloud. Around 2,000 feet altitude, something occurred that sent the MiG into an uncontrollable high-speed dive and to crash, killing both Yuri Gagarin and Vladimir Seryogin.

Theories – some conspiracy and some reasoned – abound for what happened, but no-one will ever know for certain. Perhaps the most plausible explanation is that attributed to someone who both knew Yuri and about flying: Alexei Leonov, a fellow cosmonaut and military pilot, who was on duty nearby when the crash happened. In 2013, Leonov claimed to have gained access to previously suppressed information that indicated the MiG was inadvertently 'buzzed' very closely by a new military jet – a Sukhoi Su-15 moving at or near the speed of sound, possibly as close as 50 feet or less. Turbulence set up by such a near miss would be more than enough to pitch the MiG into uncontrolled flight. Leonov studied and tested his theory for many years – even claiming to have found and interviewed the Su-15 pilot involved – and is convinced this was the cause of the crash. The price for access to the information was that the Su-15 pilot's name – he was by then in his 80s – was not to be revealed to the public.[1]

The legendary Yuri Gagarin was dead. The rulers of the USSR honoured him with a period of national mourning; a state funeral and his ashes were interred (along with those of Seryogin) in the Kremlin Wall Necropolis, a place reserved for the very highest in the land.

It was a fitting tribute to the hero who had himself flown to the highest place.

Chapter 15

Rocky Marciano

31 August 1969

orn Rocco Francis Marchegiano on 1 September 1923 in Brockton, Massachusetts, USA, the man known as The Brockton Blockbuster became a world champion heavyweight boxer and the only fighter at his weight to be undefeated during the whole of his professional career.

Having survived a serious bout of pneumonia as an infant, Rocco thrived in a close-knit Italian family, where he was brought up in a tough, multi-ethnic, working-class neighbourhood in which fists often settled disagreements at all ages. As a teenager he was very keen on baseball and took care to develop his physique, all with his eye on trying to make his way into the school baseball team – and perhaps beyond. Rocco was no mean football player, too, but baseball was his preference.

Fresh out of school, with little to offer except his brawn, Rocco went through several dead-end jobs: delivery-chute-man with the local ice and coal company – a filthy job which he detested; candy-factory worker; short-order cook in a diner and floor sweeper in a shoe factory. The war, however, boosted the construction industry and Rocco found outdoor labouring jobs easier to come by; digging holes; laying railroad track and building construction-site labouring, to name just a few jobs he had over the next two years. Soon though, the problem of finding work was resolved when, in December 1943, 'Uncle Sam' conscripted Rocco for military service in the US Army and his enlistment papers show his occupation was, at that time: 'a carpenter's helper'.

Inducted into the Army on 11 March 1943, after basic and pre-combat training at Fort Devens, Private Rocco Marchegiano was assigned to the 348th Engineer Combat Battalion at Camp Pickett. Constant training was hard work and he found it frustrating that leave-passes; girls and booze were non-existent, causing his temper and behaviour to boil over sometimes. In November 1943 the battalion was shipped to England and camped near Swansea in south Wales. By now he was a very fit young man, kept in shape by a regime of long marches, physical training, practice loading and un-loading of landing-craft and assembling and using specialist battalion heavy equipment (e.g. for bridge building). Once all this physically-demanding outdoor work was over, he worked out ways of dodging routine camp duties or even going

Absent Without Official Leave (AWOL) and sought the company of men with similar ideas.

The 348th Battalion was assigned to the invasion forces and moved down to Weymouth, Dorset prior to D-Day. Significantly, Rocco was left behind in Wales as part of the battalion reserve. On 15 June, despite strict orders to remain in camp ready to move out, Rocco and a buddy went AWOL for a last 'fling' and landed themselves in Big Trouble.

After a bout of drinking, Rocco and his pal assaulted and robbed two civilians in Filton, Bristol, beating them up badly and stealing money and personal possessions. They were arrested and court-martialled. On 17 July, Private Marchegiano was convicted and sentenced to a dishonourable discharge; hard labour and seven years in a US Army prison – later reduced to three years. Rocco is recorded as a 'general prisoner' at Fort Jay, NY, on 7 December 1944. In March 1946, he was discharged from prison having served 22 months for his misdemeanour. However, under normal Army procedure he was allowed to volunteer to re-join the Army; have his pay re-instated and most importantly, be eligible for an Honorable Discharge – and he chose this course.

In April 1946, while at home on leave, his buddy Allie Colombo talked with Rocco about taking up boxing when he left the Army and although he still dreamed of earning his living at baseball, he liked the idea he could perhaps earn some money in the boxing game, too. Back at Fort Lewis, Rocco started to train hard and boxed as an amateur for his Base and for the Army team; winning a heavyweight bout in the August 1946 Amateur Armed Forces boxing tournament in Portland, Oregon. It was during the semi-final that he sustained a serious injury to a knuckle in his left hand. He won the bout by a knockout in the first round but in doing so, badly shattered the bone of his left forefinger. Back at Fort Lewis, Rocco went into hospital for surgery on his damaged hand. It looked as if his boxing career was doomed to finish right there and then. Had it not been for the skill of Army surgeon Captain Tom Taketa, it is quite certain that given the disintegration of the knuckle, Rocco would have had no career in boxing.

At this point in Rocco's story, it is only fair to him to clarify the outcome of his service with the US Army. Rocco Marchegiano was discharged from the US Army on 27 December 1946. His US Army 'Enlisted Record and Report of Separation' document clearly states his departure was an 'Honorable Discharge.' He had paid for his mistakes and now his life moved on towards a point where boxing began to play a greater role – but not until he had let go of baseball.

While working as a construction labourer for the Brockton Gas Company, in March 1947 Rocco had a trial with Chicago Cubs baseball team. Although a big hitter, he did not make the grade – because, ironically, he couldn't make a decent throw with his right arm! It was a blow to his ambitions and he turned to his pal Allie Colombo to get him into boxing as an alternative. It was the real beginning of his future career and in the years that followed, Allie organised his training; Charley Goldman took on the role of coaching him in the dark arts of ring-craft and most significantly, helped turn his rough-house style into an unbeatable technique. Marty Weill acted as his manager at first, but was really just a front-man for his stepfather, Al Weill who was a very influential New York promoter. Soon, Rocky's potential brought Al Weill out of the shadows to take over his management, but it was never a happy relationship. Rocky had a feeling that Al was taking him for a ride on financial matters and although it did not affect his performance in the ring, it took him a long time to sort it out.

Allie fixed Rocco up with his first professional fight on 17 March 1947 in Holyoke, Mass. Fighting for a purse of $35 plus the price of his boxing licence; under an assumed name (Rocky Mack) to keep his amateur status intact. It was a four-round heavyweight contest against local boy, Lee Epperson. Rocco won by a stunning knockout in the third round and Epperson never fought again. Rocco continued boxing publicly as an amateur for another year and is believed to have had thirteen bouts in total, winning nine (seven inside the distance) and losing four; his final amateur fight coming in March 1948 when he won the New England region Golden Gloves AAU Heavyweight championship.

Rocco Marchegiano's professional boxing career continued with fight number two on 12 July 1948 at the Rhode Island Auditorium in Providence. He stepped into the ring against Harry Bilazarian, a former Army light-heavyweight champion. Only a minute into the first round, Rocco caught him with a right cross that sent him to the canvas. Bilazarian rose on the count of nine but was a sitting duck and Rocco dropped him with a right hook. The referee did not even bother to count. The audience numbered 1,000 and Rocco earned $40.

It was at this point that he and his team changed his public name, when a fight announcer was unable to pronounce his real name correctly. Rocco's manager solved the problem by introducing him to the audience as Rocky Marciano.

In his fourth fight on 9 August, Rocky was pitched in against a promising youngster, Bobby Quinn who had fifteen wins out of sixteen bouts. Quinn ran straight into a right punch that came up from Rocco's knee and he went down pole-axed, knocked out in 22 seconds of the third round. By the spring of 1950, Rocky had beaten his first twenty-five opponents, many by knockouts in early rounds, gaining him a reputation as one of the hardest hitters around. His style remained crude and awkward and he shipped a lot of punches in return but, when he landed his own, it was 'goodnight' for the guy on the other end. The wider boxing world was sitting up and taking notice.

His rapid success was all the more puzzling because, at first sight, his physical statistics were unimpressive. He was 25 years old when he turned professional – with little amateur experience – and that was considered old to start out in the fight game. At 5 feet 10½ inches in height, he was not particularly tall for a heavyweight and his reach of 68 inches was – frankly – short, since he frequently fought taller men, many of whom had a 6 inch longer reach than his own. His fighting weight, at around 185-190lbs, was again considered light for a heavyweight. By twenty-first century standards, among the 6 feet, 6 inch contenders for the four main world title belts – some of whom weigh 40lbs to 50lbs more – who would give Rocky a chance? But then, in his hey-day, who would be rash enough to bet against the little guy?

Rocky's secret lay in his supreme fitness and stamina and he feared no-one. Training for a big fight, where some boxers would spar 150 rounds, Rocky would put in 250 rounds. He developed muscle strength in his arms that enabled him to punch with great power in both hands and do it almost unceasingly – and it should be remembered that title bouts were 15-round events in those days. That destructive power is graphically displayed in iconic press photographs taken, for example, of his thirteenth round KO of Jersey Joe Walcott; the eleventh round KO of Roland LaStarza and his nineth round KO of Archie Moore.

Rocky fought in forty-nine professional heavyweight bouts and won every one, a feat not yet surpassed by a heavyweight boxer. Of these, he won eleven contests with round-one knockouts and a total of forty-three wins by KO or stoppage (TKO), the remainder went the distance and he won those on points.

On 23 September 1952, Rocky finally got his chance for a crack at the world heavyweight title, held by Jersey Joe Walcott. Despite being knocked to the canvas in the first round and going well behind on points, Rocky won the bout with a sensational KO in the thirteenth round. On 15 May 1953, he took on Walcott in a re-match and knocked him out in the first round. Walcott, however, took prize money of $250,000 to Rocky's $166,000 (≈ $1.9 million in 2024), a deal that just added another level to the anger and distrust he had about his manager. It was also another underlying reason why, in later life, he had an obsession about not only being paid a full and just reward for his efforts but being paid in wads of cash that he could actually feel, safe in his own two hands.

Rocky went on to defend the heavyweight title successfully five more times over the next three years and by his own admission, the toughest fights were with Ezzard Charles, the 'Cincinatti Cobra,' who took him all the way to fifteen rounds on 17 June 1954; Rocky winning a unanimous points decision. They met again on 17 September 1954, with Rocky coming close to suffering his first stoppage, when an Ezzard blow cut his nose very badly indeed. Split in two, it was a horrible sight, with blood spurting from the 1-inch gash. His cornerman, Freddie Brown, patched it up with thromboplastin, a quick-setting plastic gel, a swab and loads of vaseline. Rocky said it did not hurt and crucially, the ring doctor did not stop the fight. On borrowed time, Rocky chased Charles all over the ring, both men trading heavy punches. One of these opened an old cut over Rocky's left eye and he knew it was now or never. In the eighth, Rocky caught Charles with a stunning right hand that dropped him to his knees. Dazed, Charles rose to walk into a left-right combination that put him down for the full count. Rocky was out of circulation for a time while having plastic surgery on the split nose, but it was successful and never troubled him again.

Rocky's next defence was against the British heavyweight, Donald John (Don) Cockell on 16 May1955. Cockell was despatched by a KO in the ninth round, although Rocky admitted afterwards that he was a tough cookie. Following this contest, there was some public controversy about a missing $10,000 from the purse split between the boxers. It was alleged that Rocky's manager was the beneficiary of a dubious deal and this issue further soured relations between him and his manager.

Rocky's final defence of his world title came on 21 September 1955 against the wily campaigner Archie Moore, at the Yankee Stadium in New York. In his long career, Moore had fought 176 bouts, with 149 wins; 19 defeats and 8 draws and was a very skilled boxer, with a hefty punch. This was all too evident when, at the start of the second round, Moore dropped Rocky to the canvas for only the second time in his career. Under Rocky's relentless onslaught, at the end of the eighth Moore was knocked down and with his right eye closing, was saved by the bell. Both men come out strong in the ninth, but Rocky's brutal, hustling style was just too much for Moore and he finally went down for the full count.

Once Rocky had captured the heavyweight title, his earnings from boxing increased substantially. For example, his first fight with Jersey Joe Walcott reputedly earned him $94,000 ($\approx$ $1 million in 2024), while his purse for his final title defence against Archie Moore was said to be $482,000 ($\approx$ $5.6 million in 2024). Rocky's earnings in the ring and his decision to quit on 27 April 1956 while he was at the top, gave him the prospect of a comfortable life – and he was still only 32 years of age.

He was now one of the most famous people in the world, mixing with kings and presidents and in the years following his retirement from the ring, Rocky carved out a living in public relations, personal appearances, radio commentating, business ventures and even the occasional bit-part in films. To his great satisfaction, he had also rid himself of the money-draining influence of Al Weill. Always described as a mild-mannered, softly-spoken, even slightly shy man in public, Rocky gave 'spending a lot more time with [his] family,' as another reason for his decision to retire. He first met his future wife, Barbara Mae Cousins in 1947 in their home town of Brockton and they married on 30 December 1950. In 1957, they moved into a house on the outskirts of Fort Lauderdale, Florida, where they raised two children, Mary Anne and Rocco Kevin.

In 1961 Rocky had developed his public persona so well that for a couple of years he fronted his own syndicated TV series, called *Main Event*. The format of each half-hour show was for Rocky to interview a well-known guest and intersperse the chat with film clips from past boxing matches – not his own. It was all quite light-hearted and Rocky came across as lucid, articulate and well-informed about both his guest and the bout they broadcast. His guest list over the two series included Zsa-Zsa Gabor; Nat 'King' Cole; Jackie Gleason; Jimmy Durante; Tony Bennett; Joe DiMaggio; Sammy Davis Jr; Paul Newman and many others. He always presented his guest with a tiny pair of gilt boxing-gloves, in the form of a brooch, as a parting gift. Rocky appeared as a guest on many of the US prime-time talk shows of that era. These included the *Donald O'Connor Show*; *Bob Hope Show*; *Tonight, with Johnny Carson*; *Red Skelton Hour* and the *Ed Sullivan Show*. He also appeared as a guest on British TV's *Eamonn Andrews Show* on 6 November 1965, together with film actress Juliette Greco and Acker Bilk and his Paramount Jazz Band. He was never going to be the world's best actor, but Rocky had roles as a deputy sheriff in the 1960 film: *College Confidential*; as a soldier in the TV war-series: *Combat!* and there he was, in a TV and Life magazine advert for after-shave! As far as Rocky was concerned it was all enjoyable and more important to him, it brought him a tidy income – but it had to be in hard cash. He only dealt in cash and is reputed to have carried copious amounts around with him, even becoming a loan-agent in his own right, taking risks dispensing very substantial amounts as unsecured loans in the main, but the 'business' generated a good return for him – and who was going to mess with The Champ anyway? When he moved to Florida, he took to playing golf and invested in many ventures, including a bowling complex in a shopping mall in West Hollywood, Florida, the opening of which was attended by Frank Sinatra, no less.

In late-August 1969 Rocky travelled to Chicago to film a TV commercial and to discuss the contract of an up-and-coming boxer with promoter Ben Trotta. Having been entertained at the home of Andy Granatelli, head of STP, it was now Sunday, 31 August 1969, the day before his forty-sixth birthday. Rocky had finished his business in Chicago and intended to fly home to Fort Lauderdale, Florida to celebrate his birthday with his family. In his pocket he even had a freebie airline ticket from Chicago to Fort Lauderdale. However, as the result of a spur-of-the-moment decision, typical of Rocky, before he flew to Florida, he agreed to make an appearance at the launch of a new insurance business in Des Moines, Iowa. Even though there was a $500 fee involved, Rocky made this change of plan only as a favour to an old pal, the uncle of Frankie Farrell, an insurance salesman who lived in Des Moines, who was also a long-time friend of Rocky's. Frankie had persuaded his uncle to put in a word with the champ on his behalf. Rocky was persuaded to make the trip by air; someone Farrell knew had an aeroplane; it would not take up much time and Rocky could just as easily fly home from Des Moines international airport. It would all work out fine – wouldn't it? Rocky made a habit of riding small aeroplanes and had a few narrow escapes in his time.

Frankie Farrell and Rocky drove out to Chicago's Midway airport where they met their pilot Glenn Eugene Belz. The aeroplane to be used was a single-engine, four-seat Cessna 172H, registration N3149X, owned by his flying club, Executive Flyers Inc., and Glenn flew it in from Des Moines the previous afternoon to pick up Rocky and Frankie. The return flight, due west to Des Moines, should take just under three hours. Take-off was to be at 18.00 and as it was still summer, the flight should be completed in daylight. Glenn Belz was a Des Moines businessman with a private pilot licence (PPL) and 230 hours in his logbook, of which 107 were on the Cessna

172 type. He had 35 hours of night-flying logged but he was not licenced to fly on instruments alone and taken overall, he would be regarded as an inexperienced pilot.

Glenn obtained a weather briefing by telephone and was warned of a storm front towards the end of his track, with the possibility of a ceiling down to 1,000 feet or less and visibility dropping to less than 2 miles. The pilot would be aware that visual flight rules (VFR) flying, as permitted by his licence, required minimums of 1,000 feet ceiling and 3 miles visibility, but he decided to fly the trip anyway.

The Cessna took off at 5.58 pm, with Rocky sitting up front in the right-hand seat alongside the pilot and Frankie behind them. At 8.50 pm, Glenn announced his approach by radio to Des Moines air traffic control, advising them that he could not see the city due to poor visibility in cloud and drizzle. Requesting a radar 'fix,' it located him approaching the town of Newton, Ohio and reporting that he could see the lights of Newton, he said he would land at the municipal airport to refuel. At 9.05 pm, in fading light, the aeroplane was seen, from the ground, to come out of low cloud at about 100 feet then disappear back into the cloud. It dropped below cloud again, banking and heading towards the ground. Suddenly, the Cessna's starboard wing hit a tree, shearing it off. The aeroplane careered across a field, disintegrating catastrophically and striking another tree before coming to rest in the bed of a dry creek on farmer Henry Eilander's land. No-one survived. Belz was flung out and found beneath the engine. Farrell, too, was found dead outside the wreckage. Rocky was still in what was left of the cabin but his skull had been fatally pierced by a shard of metal wreckage.

The National Transportation Safety Board (NTSB) investigation concluded the probable causes of the accident were: that the pilot attempted an operation beyond his experience level; that he continued a VFR flight into adverse weather conditions and suffered spatial disorientation that caused an uncontrolled descent into the ground, 1.5 miles south-west of Newton airport.

Reading between the lines, though, one can detect an element of pressure to have Rocky make an appearance at a dinner in the Charcoal Rooms, Des Moines to launch Frankie Farrell's new business. It has been suggested that it would be a surprise party for Rocky's birthday, too. It is most unlikely that that pressure would come from Rocky. He would be unaware that friends were gathered at Frankie's house for the surprise party; his speech, in support of Frankie, was something he was very used to doing and it is unlikely he would have even worried much about a cancellation due to weather, because that would mean he could set off for Florida to be with his family. So, if any pressure to make the trip existed, it would most likely have been felt by Frankie and maybe some of that transferred to the pilot.

Across the world, Rocky will never be forgotten and in Brockton, a statue of this boxing legend stands in the grounds of the Rocky Marciano Sports Stadium. He wrote himself into the annals of world boxing and thus will always appear in any discourse of that noble sport's history as the Undefeated Heavyweight Champion of the World.

Chapter 16

Audie Murphy

28 May 1971

Congressional Medal of Honor recipient, Audie Murphy once remarked:

> 'I've been fed up with the accolade of America's most decorated soldier for a long time. I gave away most of my medals because I felt they never belonged entirely to me. My whole unit earned them, but I didn't know how to give them to my whole unit.'

For all his self-effacing honesty, when Audie Leon Murphy's life ended abruptly on the rain-lashed slopes of Brushy Mountain, Virginia, that well-deserved accolade would continue to shine brightly for ever. Audie was 45 years of age when he died on 28 May 1971 but, in that short life, he travelled from poverty to riches; had been – as reflected by the title of his autobiography – *To Hell and Back* and had changed his war-hero persona into a movie and TV star; poet; song-writer; rancher and businessman.

The son of a cotton share-cropper: Emmett Berry Murphy and his wife Josie Bell, neé Killian, Audie, one of eleven children, was born on a farm near Kingston, in Hunt County, Texas on 20 June 1925. After the family moved to Celeste, Texas, in 1933, his father walked out when Audie was twelve, whereupon his mother, together with three of her youngest children, Joe, Billie and Nadine, moved into the home of a married daughter, Corrine Burns, in Farmersville, Texas. To help support his mother and the children, Audie left school in the eighth grade to work as a farm labourer and when farm work dried up, he took a clerk/handy-man's job in a grocery store and petrol station. He was 16-years-old when his mother died at the age of 50 and this sad event caused the break-up of the young family. With her mother gone, Corinne could not cope with her own children and keep on the other three, so Joe, Billie and Nadine were placed in Boles Orphanage Home, close by.

Audie went to nearby Greenville, Texas where he rented a room and found a job in a radio repair shop and was working there in December 1941 when the USA entered the Second World War. Audie's birth records were destroyed in a fire at a registry office, so he needed a new birth certificate to apply for service in the armed forces. In mid-1942 and keen to join the US Marine Corps Audie, then still a skinny kid of only 17 years of age, conspired with a local doctor – who sponsored him – to falsify his birth year to 1924, so that he could present a birth certificate at enlistment which showed him to be aged eighteen. One of Audie Murphy's abiding and most charming physical attributes was his open, freckled, baby-faced visage – something he never really lost – but the Marine Corps turned him down as being too short (5 feet 5½ inches) and under-weight (112lbs). Still keen to join up, he now applied to the US Army, which rejected him for his first choice as a parachute soldier but finally accepted him into the infantry on 29 June 1942. As it turned out, he was destined to become one of the finest infantry soldiers in the US Army.

As a raw recruit, 18083707 Private Audie Murphy was called up to Fort Wolters, Texas in October 1942 for basic military training, then on to Fort Meade, Maryland for advanced infantry instruction. Gaining high marks for attitude; marksmanship and leadership, he was posted on 23 January 1943 to Camp Kilmer, New Jersey, to await a ship to Morocco in North Africa. He sailed on 8 February from New York as a replacement assigned to the 15th Infantry Regiment. In Casablanca he joined 'B' Company, 1st Battalion, 15th Regiment, 3rd Infantry Division but saw no combat during the tail-end of the Tunisian campaign that finally cleared Axis forces from North Africa. Unusually and despite periods in hospital recovering from wounds; the effects of malaria and even when commissioned later, he remained with 'B' Company throughout his war service.

By the time the war was over in Europe, in little more than two years, Audie had participated in seven major Allied campaigns: Sicily, Naples-Foggia, Anzio, Rome-Arno, Southern France, Rhineland and Ardennes-Alsace. During this time, he was awarded no less than twenty-four US decorations and campaign medals, including a Congressional Medal of Honor; Distinguished Service Cross; Silver Star with an Oak Leaf Cluster; Legion of Merit; Bronze Star; Purple Heart with two Oak Leaf Clusters and the European/Africa/Middle East campaign medal with seven battle stars, and many other awards. His foreign decorations include a Chevalier of the Legion d'Honneur and the Croix de Guerre with Silver Star (France) and a Croix de Guerre with Palm from Belgium.

On 15 August 1944, American, British, Canadian and Free French forces mounted a sea-borne invasion: Operation *Dragoon*, on the Riviera coast of the south of France. While pinned down and taking casualties in the vicinity of La Ferme du Vallon, near Des Bouin, Audie's extraordinary heroism resulted in the capture of a fiercely contested enemy-held hill and the annihilation or capture of its enemy garrison; earning him a Distinguished Service Cross.

As the regiment pushed enemy forces ever northwards through France, Audie put himself in such exposed positions that it seemed a miracle his life was not snuffed out in an instant. Following the enemy's retreat along the Rhone valley, Murphy's regiment met stiff resistance near the city of Montelimar, but the enemy's retreat continued towards the Moselle and the Vosges mountains. At dawn on 2 October, Audie together with 1st Battalion commander, Colonel Paulick; his operations officer, Colonel Ware; Captain Harris, 'B' company commander and three enlisted men, were on a reconnaissance patrol near Cleurie Quarry. Moving across the scrubby hillside, they came under machine gun fire from a concealed German position. Two men fell wounded; the rest dived for cover and a fire-fight ensued. There was no way back unless that machine gun could be neutralised. Disregarding enemy fire, Audie charged and flung two hand grenades at the machine gun and destroyed it. It brought him a Silver Star. With the enemy in retreat, in the afternoon of 5 October 1944, near Le Tholy, Audie helped to hasten their departure. Carrying a portable radio, he crawled fifty yards under severe machine gun and rifle fire, to a point 200 yards from the entrenched enemy. Despite a hail of bullets that hit as close as a foot to him, Audie directed artillery fire upon enemy positions for an hour, ensuring the retreat continued. This selfless action brought him a second Silver Star.

Audie Murphy was offered a battlefield commission on 15 October 1944. His regiment had lost many officers and having resolutely resisted a battlefield commission – on the grounds that Army rules meant his acceptance required him to transfer to another Unit, which he absolutely did not want – he finally gave in, because his regimental commander, as a matter of local expediency, did not enforce the transfer rule. That was how 01692509 Second Lieutenant Audie Murphy managed to stay with his men as their new platoon commander.

Tasked with capturing Brouvelieures in the French Vosges region, on 26 October, his 3rd Platoon came under enemy sniper fire and Audie took a bullet through his hip. In appalling weather that turned roads into a quagmire, his ragged wound had turned gangrenous by the time he reached US 3rd General Hospital in Aix-en-Provence. He had surgery on the infected wound and was medicated with the wonderful new penicillin drug but Audie was out of action for ten weeks, re-joining his company at the front line on 14 January 1945.

On 24 January, the ferocious battle of the Colmar Pocket began. Fought in bitter winter conditions, with several feet of snow covering rock-hard frozen ground, it was here that Audie put his life on the line yet again. On 25 January Audie was slightly wounded in the left leg by shrapnel from a mortar blast, but he refused to be evacuated. Responding to American attacks, the Germans mounted a dogged defence, using tank-supported infantry counter-attacks, aimed at re-capturing nearby woods. Casualties had depleted Audie's company strength by forty percent and he was the only officer left. In the early hours of 26 January, he was ordered to take command of 'B' Company and hold his sector of the Riedwihr Woods. That afternoon, the Germans mounted a surprise attack from the Holtzwihr direction towards his thinly-manned line at the edge of the wood. If they broke through it would endanger the 3rd Division's entire position. The following medal citation tells the story of what happened that day.

John Alcock stowing a coffee flask aboard the Vickers Vimy prior to his trans-Atlantic flight with Arthur Whitten Brown, on 14 June 1919. (Public domain)

Vickers Viking Amphibian prototype, G-EAOV in which Sir John Alcock lost his life near Rouen in December 1919. (Public domain)

Roald Amundsen leaning against a Dornier Wal flying boat at Svalbard while preparing for his attempt to reach the North Pole by air. (Courtesy of National Library of Norway, public domain)

The Latham 47, No. 02, flying-boat, at Bergen in June 1928; used by Amundsen in his search for the lost airship *Italia* and the aeroplane in which he lost his life. (Courtesy of National Library of Norway, public domain)

Juan de la Cierva (right) in his Cierva C-8L autogiro, registration: G-EBYY; a Lynx-engine Avro 504 fuselage. Here he is at Le Bourget airport, after making the first international flight by an autogiro, from London to Paris, on 18 September 1928. His passenger (left) is Mons Henri Bouché, editor of *La Revue d'Aeronautique*. (Bibliotheque nationale de France; catalogue.bnf/fr/ark:/12148, public domain)

Three KLM Douglas DC-2s line-up at Amsterdam airport in the late-1920s. On the left is PH-AKL, with the name *Lijster* (Thrush) on its nose. The other two are, centre, *Toeken* (Toucan) and right, *Patrijs* (Partridge). Juan de la Cierva was a passenger and died when *Lijster* crashed at Croydon on 9 December 1936. (Courtesy of Peter's blog, www.patmcast.blogspot.com_2012_01_teneningen-2, Creative Commons Attribution-Share Alike 3.0)

Left: Amelia Earhart and Lockheed Vega 5C, NR965Y, at Burbank, California in December 1934, prior to her flight from Hawaii to California on 11 January 1935. (Courtesy of Robert S. Ball Special Collection, Wright University, USA)

Below: Lockheed Electra 10-E, NR16020, in which Amelia Earhart and Fred Noonan were lost during their round-the-world flight. (W.T. Larkins Memorial Collection, Courtesy of Johan Visschedijk – 1000aircraftphotos.com)

Dressed for the heat, Amy Johnson and her DH60 Moth *Jason* take a well-earned break in Jhansi, India, *en route* to Calcutta (Kolkata), 11 May 1930 on day-8 of her epic flight from England to Australia. (Courtesy of Wikipedia user: 'Dabbler,' under GNU Free documentation licence version 1.2)

Airspeed Oxford II of the type in which Amy Johnson made her final flight. (Courtesy of Alan Brown collection, under GNU Free documentation licence version 1.2)

Above: Douglas DC-3 of Transcontinental & Western Airways (TWA), representative of the aircraft in which Carole Lombard lost her life on 16 January 1942. (Courtesy of The Finnish Heritage Agency, museovirasto.fi. Creative Commons Attribution 4.0 International licence)

Left: British actor Leslie Howard, left, with Norma Shearer in the film *Romeo and Juliet*, made in Hollywood in 1936. (Courtesy of Folger Shakespeare Library Digital Image Collection, under Creative Commons Attribution-Share Alike 4.0 International licence)

Above: KLM Douglas DC-3-194 named *IBIS*, at Malmi airfield, Helsinki on 1 July 1939; the aeroplane in which Leslie Howard lost his life. (Courtesy of The Finnish Heritage Agency, museovirasto.fi. Ref: JOKAHBL3B_J30:1. Creative Commons Attribution 4.0 International licence)

Right: Glenn Miller at the height of his fame in 1942. (Public domain)

A Noorduyn UC-64A Norseman aircraft, 44-70370, representative of the USAAF Norseman 44-70285 in which Glenn Miller was lost over the English Channel on 15 December 1944. (Courtesy of Bill Larkins, under Creative Commons Attribution-Share Alike 2.0 Generic licence)

Left: A rare passport photograph of Roy Chadwick age 23 in 1916. (Courtesy of Delphine Stevens)

Below: Avro Tudor 2, G-AGSU, the aircraft in which its designer Roy Chadwick was killed when it crashed on take-off on 23 August 1947. (Walter van Tilborg Memorial Collection, Courtesy of Johan Visschedijk – 1000aircraftphotos.com)

Airspeed AS.57 Ambassador Mk2, G-ALZS of British European Airways (BEA), similar to aircraft G-ALZU which crashed at Munich on 6 February 1958. (Ed Coates Collection, Courtesy of Johan Visschedijk – 1000aircraftphotos.com)

Beech 35 Bonanza, 1947 model, similar to the aircraft in which Buddy Holly lost his life. (Courtesy of Johan Visschedijk Collection – 1000aircraftphotos.com)

United Nations Secretary-General Dag Hammarskjöld (left) with Prime Minister of Israel, David Ben Gurion at Sde Boker in 1959. (Courtesy of Moshe Pridan of Israel government press office; public domain)

Douglas DC-6B, SE-BDF of Transair, Sweden is a representative aircraft for SE-BDY of Transair, in which Dag Hammarskjöld lost his life on 17 September 1961. (Courtesy of Ralf Manteufel collection, Wikimedia Commons GNU Free Documentation licence, version 1.2)

Sud Aviation SE-210 Caravelle 10R, EC-BDD, *Jesus Guridi*, of Iberia, is the airliner in which actress June Thorburn lost her life on 4 November 1967. (Courtesy of Erik Frikke and airhistory.net)

Right: Russian cosmonaut Colonel Yuri Gagarin at a press conference in Helsinki on 30 June 1961. (Courtesy of The Finnish Heritage Agency, museovirasto.fi. Creative Commons Attribution 4.0 International licence)

Below: A MiG15 UTI two-seat Russian trainer aircraft of the type in which Yuri Gagarin was killed on 27 March 1968. (Courtesy of Tony Hisgett, Creative Commons Attribution 2.0 Generic licence)

A Cessna 172H Skyhawk, representing the type of aeroplane in which Rocky Marciano died on 31 August 1969. (Author's photo)

Above left: War hero and movie star, Audie Murphy. (Courtesy of Mindy Smith Collection)

Above right: Portrait of the international milliner Otto Lucas. (Courtesy of Rolf Andresen, via Dr Anna Nyburg)

An Aero Commander 680E of the type of aircraft in which Audie Murphy lost his life on 25 May 1971. (Courtesy of Jeff Gilbert, JGphotographics.com; GNU Free Documentation licence version 1.2)

Vickers 951 Vanguard, G-APEC of British European Airways, seen at London Heathrow airport in 1965, is the airliner in which Otto Lucas died on 2 October 1971. (Courtesy of Adrian Pingstone; arpingstone; public domain)

Graham Hill driving a BRM P57 '12' during practice for the 1963 Dutch Grand Prix at the Zandvoort circuit on 21 June 1963. (Courtesy of National Archives of Nederland, nationaalarchief.nl; photo by Harry Pot/Anefo, archive number 915-2837, Creative Commons Zero/CC0; public domain)

Piper PA-23-250D Aztec, N6645Y, in which Graham Hill died. Seen at Blackbushe airport before it was re-painted in Embassy-Hill team colours. (Courtesy of Peter Brown and blackbusheairport.proboards.com)

Francis Gary Powers, U–2 pilot, holding a model of a U–2, giving evidence at the US Senate Armed Services Committee hearing in Washington DC on 6 March 1962. (Courtesy of Library of Congress, ppmsca 39634; photo by Warren K. Leffler, public domain)

Above left: The Bell 206B JetRanger, N4TV (formerly N555TV), the KNBC 'Telecopter' in which pilot Francis Gary Powers and cameraman George Spears lost their lives on 1 August 1977. (Courtesy of Earlytelevision.org/archive helicopters, under licence Creative Commons Attribution–Share Alike 4.0)

Above right: American golfer Payne Stewart, in August 1998 at the Vancouver Open in Canada. (Courtesy of Ed Balaun/Supergolfdude, under Creative Commons Attribution–Share Alike 4.0 International licence)

A Learjet Model-35A business-jet, similar to N47BA in which Payne Stewart lost his life. (Courtesy of 'bomberpilot' reference number: 8738280921; Creative Commons Attribution–Share Alike 2.0 Generic licence)

Hawker Siddeley (BAe) HS748, ZS-OJU, the aeroplane in which Hansie Cronje lost his life. (Courtesy of Michel Anciaux)

The ill-fated Transaven LET L-410UVP Turbolet, YV2081, which crashed in the sea on 4 January 2008, seen here at El Gran Roque aerodrome, Venezuela on 16 Dec 2006. (Courtesy of Enriques Perrella)

Brittain-Norman Islander BN-2A, registered as N555JA, in a very stripped-down condition at an airfield in Florida, USA, before being sold to Transaereo 5074 in Venezuela and registered as YV2615. This is the aircraft in which Vittorio Missoni and his friends were lost on 4 January 2013. (Courtesy of John Bennett)

Left: Los Angeles Lakers basketball legend, Kobe Bryant in action against the Washington Wizards on 12 March 2014. (Courtesy of Keith Allison; Creative Commons Attribution-Share Alike 2.0 Generic licence)

Below: The Sikorsky S-76B helicopter, N72EX, in which Kobe Bryant lost his life on 26 January 2020. (N72EX by Don Ramey Logan.jpg from Wikimedia Commons, by Don Ramey Logan [https://don.logan.com] CC-BY-SA 3.0 [https://creativecommons.org/licenses/by-sa/3.0/deed.en)

'By direction of the President, a Medal of Honor for conspicuous gallantry and intrepidity at the risk of life above and beyond the call of duty was awarded to Second Lieutenant Audie L. Murphy, 01692509, 15th Infantry, Army of the United States, on 26 January 1945, near Holtzwihr, France, commander Company B, which was attacked by six tanks and waves of infantry. Lieutenant Murphy ordered his men to withdraw to a prepared position in a wood while he remained forward at his command post and continued to give fire directions to the artillery by telephone. Behind him to his right, one of our M-10 tank destroyers received a direct hit and began to burn. Its crew withdrew to the woods. Lieutenant Murphy continued to direct artillery fire which killed large numbers of the advancing enemy infantry. With the enemy tanks abreast of his position, Lieutenant Murphy climbed on the burning tank destroyer which was in danger of blowing up any instant and employed its .50-caliber machine gun against the enemy. He was alone and exposed to the German fire from three sides, but his deadly fire killed dozens of Germans and caused their infantry attack to waver. The enemy tanks, losing infantry support, began to fall back. For an hour the Germans tried every available weapon to eliminate Lieutenant Murphy, but he continued to hold his position and wiped out a squad which was trying to creep up unnoticed on his right flank. Germans reached as close as 10 yards only to be mowed down by his fire. He received a leg wound but ignored it and continued the single-handed fight until his ammunition was exhausted. He then made his way to his company, refused medical attention and organized the company in a counter-attack which forced the Germans to withdraw. His directing of artillery fire wiped out many of the enemy; he personally killed or wounded about 50. Lieutenant Murphy's indomitable courage and his refusal to give an inch of ground saved his company from possible encirclement and destruction and enabled it to hold the woods which had been the enemy's objective.'

Having crossed the river Rhine, 3rd Infantry Division was withdrawn and placed in reserve for an assault on the Siegfried Line. Now considered a valuable commodity Audie, too, was withdrawn from the front line, promoted to First Lieutenant on 22 February and given a liaison job at 15th Infantry Regiment headquarters.

The Congressional Medal of Honor (CMH) is the highest award for valour in military action that his country could bestow and equates to the Victoria Cross in the British Commonwealth. His CMH together with a US Legion of Merit medal, was presented by Lieutenant-General Alexander H. Patch III, commander of the US 7th Army in Europe, in a ceremony held on an airfield at Werfen, near Salzburg, Austria on 2 June 1945. Audie was just a couple of weeks short of 21-years of age – according to his 'doctored' birth certificate – a battle-hardened war veteran and national hero, but still not old enough to vote in US elections.

Such prolonged exposure to death and destruction took its mental toll and in the years following the war, Audie was plagued with insomnia, anxiety and flashbacks that eventually caused him to resort to strong medication, which then brought its own problems of addiction. It took all his indomitable spirit to wrestle with his addiction and to succeed in overcoming these personal demons.

Returning to Texas in the summer of 1945, Audie's life took a very significant turn when his picture appeared on the cover of *Life Magazine*, just before he was discharged from the Army that September.

Among the readers of the *Life Magazine* article was the film star James (Jimmy) Cagney who invited Audie to Hollywood to discuss taking up a career in the movies. After much heart-searching, he took up the offer and Cagney arranged for him to have acting, voice and dance coaching. Although contracted to Cagney's production company, no film work came Audie's way and they parted in 1947. Audie took up residence in Hollywood and got to know a few of the 'right people' who helped him to secure 'bit-parts' in a couple of films. His first starring role came in his third film: *Bad Boy* (Allied Artists: 1949). In that year, too, Audie married actress Wanda Hendrix, but they divorced after about a year. His first western movie was *The Kid from Texas* (1950), made at Universal-International Studio, which gave him a long-term contract with a pay-cheque of $2,500 a week (≈ $32,000 in 2024), setting him on a path to movie fame. He also found time in his busy filming schedule to marry Pamela Archer, eventually raising two sons, Terry Michael and James Shannon.

Credited with roles in forty-four feature films, the peak of Audie's film career came when he starred in a film based on his autobiography of the same title: *To Hell and Back* (Universal-International, 1955). This film is unique because of Audie's portrayal of himself – something not previously done on the silver screen. His share of the profits reputedly earned him $500,000 (≈ $4 million in 2024) and in addition to his income from a string of films, he invested wisely in a ranch-style house in the San Fernando Valley and later, a ranch in Arizona.

Audie Murphy is most remembered for his roles in the western film genre, which made him a great deal of money. It should not be forgotten, though, that he had roles in light comedy, such as *Joe Butterfly* (1957) and in dramas: *World in my Corner* (1957); *The Quiet American* (1958) and *Trunk to Cairo* (1967).

In the mid-1950s, Audie became an active freemason for most of the remainder of his life and indeed his Medal of Honor is displayed at the Dallas Scottish Rite Temple Museum. By 1960, Audie owned a 17,000-acre ranch in Vail, Arizona, where he kept 400 head of cattle and bred quarter-horses as part of his interest in horse-racing. He loved horses, becoming a skilled rider and preferring to ride his own horses in his many western movies. Audie enjoyed riding in quarter-horse races and gambling, too – probably a bit too much, since the latter seems to have contributed to some of his later financial difficulties.

Yet, there was a softer, more artistic side to Audie Murphy that is often overlooked but is visible through his portfolio of song lyrics and poems. During the 1960s he collaborated with Scott Turner and others, on several country and western-style songs. Among the most successful of his twenty or so songs were: *Shutters and Boards*, (1962), recorded by over fifty different singers around the world, while his last song: *Was it all Worth Losing You* (1970), was recorded by the legendary Charlie Pride.

Audie owned another house near Toluca Lake and a 60-foot motor boat and it was around this time that he took up flying as a hobby; life was pretty good. However, due to his near-addiction to gambling and a series of unsuccessful business ventures, in 1968 Audie stood on the verge of bankruptcy. By this time, too, he had virtually retired from making movies, but he was not a man to let the world get on top of him and gradually restored his finances through self-discipline and new business ventures. His name and public profile were obviously a great asset in helping him re-shape his future but he was never the sort of person to overplay his military service and decorations. By 1971 he was a director of Telestar Leisure Investments, a conglomerate of companies based in Denver, Colorado and he had also set up his own film production company: First International Planning Co., (FIPCO) which produced its first movie, a western: *A Time* for *Dying* (FIPCO: 1971), in which Audie played his final film role.

In his position as director of Telestar, Audie was always examining new business investments on behalf of his clients and he was contacted by a company called Modular Management, which was seeking to raise capital to expand one of its factories manufacturing pre-fabricated homes and other buildings. On 28 May 1971, Audie was in Atlanta, Georgia, from where he was due to fly to Martinsville, Virginia to inspect the Modular Properties Inc., manufacturing plant. Audie was accompanied on this trip by Jack Littleton, representing a group of Californian investors, together with the president of Modular Management, Claude Crosby and Raymond Pater, an attorney representing Modular Management. The final passenger was Kim Dodey, a friend of Jack Littleton.

The aeroplane in which the party would make their trip was a twin-engine, six-seat Rockwell Aero-Commander 680 E, registration N601JJ, recently acquired by Colorado Aviation Inc., a company set up by Telestar. Their pilot, was Herman Levelle Butler who, the previous day, had flown the Aero-Commander to DeKalb-Peachtree airport, on the northern outskirts of Atlanta, from its base in Denver, to pick up Audie Murphy and his companions. Butler had 8,000 flying hours as a PPL (private pilot licence) pilot; he was rated to fly single- and twin-engine aeroplanes, but he did not have a rating to fly on instruments under Instrument Flight Rules (IFR), in what are known as Instrument Meteorological Conditions (IMC). Later investigation established he had only six hours experience in flying the Aero-Commander 680.

The flight took off from DeKalb at 9.10 am, on Friday, 28 May in good weather but without a flight-plan being filed. While cruising eastwards under visual flight rules conditions (VFR), the weather along the route gradually deteriorated. With a lowering cloud base, rain and fog reduced visibility, forcing the pilot to fly ever lower to maintain visual progress along his track. In the vicinity of Galax, a town in south-west Virginia, the aeroplane is thought to have encountered the overcast and it is considered likely that at this point the pilot began to lose his visual references and would have had to resort to flying on instruments.

From later reports made by some residents of Galax, at around 11.00 am, it is believed the pilot made several low-level circuits of the town and local area apparently seeking a place to land. There was an airfield at Hillsville, north of the town and perhaps Herman Butler was aware of its existence but was unable to locate it. At 11.49 am, still flying in the area at low-level, Butler made radio contact with the airport at Roanoake, 60 miles north-east of Galax, for a weather report and indicated his intention to land at that airport. Flying in the overcast and now probably unsure of his location, the aeroplane was getting perilously close to, or maybe even already among the mountains to the west of his track.

It was shortly after 12.00 noon that the Aero-Commander made its fatal impact 2,700 feet up the north side of Brushy Mountain, at a point roughly 300 feet below its summit; about 20 miles west of Roanoake. The National Transport Safety Board (NTSB) report of June 1972 stated that at the time of the accident, visibility at the impact site was zero, with fog and rain and it categorised the weather conditions as IFR. The conclusion of the NTSB was that the probable cause of the accident was due to the pilot continuing to take his visual flight into adverse weather conditions, at an altitude too low to clear mountainous terrain and the pilot failing in his attempt to operate the aeroplane beyond his experience and ability level, in instrument conditions. It was considered that everyone on board the aeroplane had died instantly.

In the absence of a flight plan and because no-one actually witnessed the crash, it was not until the afternoon of Monday 31 May that the wreckage was discovered during a wide-scale air search. Thirty-one aeroplanes from the Virginia Civil Air Patrol (CAP) airfields at Martinsville, Lynchburg, Roanoake, Danville, Hillsville and Buckingham took part in the search operation. Rescuers were then directed on the ground and hiked four miles up the mountain to reach the site. Although the rear fuselage and tail-fin was damaged but intact, they found the forward fuselage and cabin was completely destroyed and there was evidence of a post-impact fire. All six bodies were recovered and taken down the mountain by four-wheel drive vehicles. One piece of documentation used to identify the bodies was Audie's drivers licence, which had printed on it his real date of birth: 6-20-25 (20 June 1925).

For all Audie's fame, broadcast media coverage of the passing of this national hero was pitifully thin, adding up to just a few minutes air-time in total in news bulletins across the country. This was, however, made up for by his burial with full military honours in Arlington National Cemetery on 7 June 1971 and the numerous press articles and memorials that emerged after his death. In 1991, for example, the Appalachian Trail was re-routed so that its track would pass by a granite stone memorial to Audie Murphy, placed on top of Brushy Mountain close to the crash site. In 1960 he was honoured with a star set in the pavement on the 'Hollywood Walk of Fame' in Vine Street, Hollywood, California. And, not least, there are his trademark western films which, in the 2020s, are shown virtually every week on British television, thus reminding audiences of Audie Leon Murphy and the contribution he made to American history.

Chapter 17

Otto Lucas

'Bealine seven-oh-six; we're going down; we're going down! Mayday! Mayday! Mayday! We're going down vertically! Out of control! Out of control!' This was the last radio message from the crew of Vickers Vanguard G-APEC to Brussels Air Traffic Control. On the morning of 2 October 1971, in the space of five minutes, British European Airways (BEA) flight 706 was transformed from cruising along serenely at 19,000 feet over the Belgian countryside, to plummeting suddenly to earth in a dive that killed all sixty-three souls on board.

'If you want to get ahead, get a hat,' goes the old saying. Step forward Otto Lucas, who certainly put that into practise to became one of the foremost British milliners of the mid-20th century.

Lucas' childhood and early family life is unrecorded, while his rise to prominence in his profession was only detected and then increasingly reported upon by the world press, after the Second World War. There is some suggestion that he 'learned his trade in Berlin and Paris' and 'his arrival in England [was] in 1932, when he opened business premises in Bond Street, London.' Although there is no primary evidence to support the Berlin or Paris claims, London magistrates' court evidence in January 1933 confirms that he had a millinery business in Old Cavendish Street at that time, possibly established with initial funding from the wealthy Jewish philanthropist Louis (Sir Louis; 1937) Stirling (1879-1958).

While in his heyday Otto Lucas projected an air of flamboyancy, he led a discrete private life. This might be explained by his personal and private life during a period of history where being German, Jewish and gay, were socially and politically problematic. Otto was a member of a notorious private drinking club called The Colony Room, in Dean Street, Soho, where he was very friendly with its first proprietor, the outrageous bohemian, Muriel Belcher. In the decade preceding his death, Otto's male companion and live-in partner was a Norwegian by the name of Rolf Andresen.[1]

Otto was born in Mulheim, Germany on 9 July 1903, to Jewish parents Jacob and Dina Sara Lucas (neé Mannheimer). He had an elder sister, Erna (born 10 June 1902), who married a fellow-Jew, Julius Katten (born Frankenburg, 14 May 1895) and they migrated to London in the 1930s as refugees, taking up residence at 57 Compayne

Gardens, NW6. This address is shown on Julius's Internee Record card in 1939 and also on his naturalisation application in 1948. It is possible that the reason for Otto taking up residence in London was to escape the growing antisemitic activity of the Nazi party in Germany and the timings support that assumption. There is evidence from Dutch Jewish Holocaust records that Otto's parents left Germany for Holland in the years immediately before the Second World War, possibly for similar reasons.

A letter, written on 12 July 1943 by Otto Lucas to Pope Pius XII has come to light, in which he pleads for the Pope's help to trace and save his parents from the hands of the Nazis. It shows that by that time Otto was living at 48 Berkeley Square, London – another prestigious residential address.

Sadly, Otto's plea failed to save his parents, since they were already in Auschwitz and were doomed to share the fate of millions like them. Having taken up residence in Amsterdam and being Jewish, they were swept up in March 1943 during the German occupation of Holland and deported, via the Dutch Nazi party camp at Westerbork, to Auschwitz concentration camp, where Holocaust records indicate they died in that camp on or about 17 September 1943.

It is during the 1930s that Otto Lucas established himself as a major player in the British millinery trade; to such an extent that his success can perhaps be measured by the quality of his home addresses. On 7 July 1938, for example, *The Times* (London) carried a notice that Otto had applied for naturalization as a British citizen. His address at this date was given as 61 Cadogan Square SW1, a grand apartment building of character in Mayfair. It seems that his naturalisation application was unsuccessful, possibly due to the delicate political situation and the onset of war having delayed the process. In Britain, around 72,000 enemy aliens were identified in 1939 and subjected to a security investigation. When war was declared these 'enemy aliens' had to register their details with the police. Each was then assessed for their potential threat to national security at one of 120 tribunals set up across the UK.

When the internment tribunal process began in September 1939, Otto Lucas was picked up by the system and was registered with the police as case number 524326; living at 41 Gilston Road, West Brompton, Kensington & Chelsea, London – a fine building located in a select district. His 'Male Enemy Alien – Exemption from Internment – Refugee' record card shows his occupation as 'Millinery – Wholesale' and that he owned his own business at 87 New Bond Street, London, confirming that he had moved his business to new – probably larger – premises. At this time, it is believed Otto was only required to report daily to his local police. The tribunal decision, dated 1 December 1939, was that he was: 'not to be interned.' However, having been allowed his freedom, this was reversed in 1940 – for Otto and tens of thousands of others in a similar position. Otto's card is stamped: 'INTERNED CIRCULAR 21.6.40,' which restricted his movements again. There is no mention on his record card of him being sent to a camp, although the *Association of Jewish*

Refugees Journal (November 1971) states that: 'in the summer of 1940, Otto Lucas was a spokesman of the internees who were detained in Lingfield at this time'. This refers to the internment camp set up on Lingfield race course. Fortunately, Otto's new review was favourable and his record card is endorsed: 'Release Authorised. Released 9.9.40, Category 3.'

During the 1930s, Otto is known to have struck millinery deals with David Jones Limited stores in Australia and New Zealand, in addition to his UK and American outlets.

Otto was a wholesaler and his business method was to buy model hats that he felt had mass-market potential, from all the Parisian couture houses, such as Maude et Nano, Orcel, Legrou, Dior and Lanvin. He would take them back to his own designers in London, who would work out how to produce something similar to the original but with his own twist. The bulk of his wholesale business was aimed at stock-buyers from all the well-known department stores in the UK and overseas. Buyers from Harrods; Fortnum & Mason; Debenham & Freebody; Harvey Nichols; Jenners of Edinburgh; McDonalds: Glasgow; Samuels: Manchester; Pearsons: Huddersfield; Griffin & Spalding: Nottingham; Dennett's of Wilmslow and Saks of Fifth Avenue and Lord & Taylor of New York to name but a few, would visit the Bond Street premises to inspect and barter with the sales force. Similar bulk business was conducted with stores in major European cities, Australia and the USA. Post-WW2, Otto also built up a large bespoke-millinery business dealing with private clients from among the wealthy, the nobility and celebrities of stage; screen – and everywhere else. He could count such personalities as Princess Marina, Duchess of Kent; Wallis Simpson, Duchess of Windsor; Princess Anne; Princess Lee Radziwill, sister of Jackie Kennedy and the actress Greta Garbo in his portfolio. In his time, examples of Otto Lucas in-house hat designs appear on the front cover of no less than twelve different issues of the iconic *Vogue* (UK) fashion magazine, between 1955 and 1968; his creations being captured by some of the leading photographers of the day, including David Bailey, Brian Duffy, Barry Lategan, Helmut Newton and Claude Virgin.

Another highly successful and future Royal milliner, Philip Somerville, worked for Otto Lucas during the mid-1960s and recalled:

'Otto Lucas was brilliant; he made the most beautiful hats. His philosophy was that every hat should suit a hundred people, it was not individual hats for individual people. A person should come into a hat shop and be able to try on ten hats or twenty hats and look good in all of them. He was also strict about hats being uncomplicated. One of the first things he said to me was: "Ah, Somerville, what do we take off that hat?" And we would take off a flower here, or a bow off there. He was a very classic, uncluttered person.'

Somerville also remembered that around 1968, the wholesale price of an Otto Lucas hat, for the mass-market (dubbed: 'shopping hats') and sold to the likes of Debenham & Freebody, was £30 (≈ £440 in 2024). He said such a hat could be sold retail in these up-market stores for around £75 (≈ £1,100 in 2024). To give some financial perspective to the size of the business, it is said that in his final year, Otto's company produced 55,000 hats for the wholesale market. If a price of £30 is used. that would represent a gross turnover of £1,650,000 (≈ £21,000,000 in 2024).

When the Second World War ended, there was still a buoyant market for fashion and daily hat-wearing by men and women was still prevalent at this time. The Otto Lucas company's reputation ensured it was well-positioned to take advantage of the female desire to cast off the drabness of austerity. June 1947, for example, saw one of the leading British ready-to-wear fashion companies: Brenner Sports, hold its Autumn Collection show in London, at which all the hats were supplied by Otto Lucas Ltd. *The New York Times* declared on 5 July 1950 that 'In the Fall, the millinery of Otto Lucas of London will be widely available at Saks of Fifth Avenue.'

During 1951/52, in an effort to encourage home industry, its own exports and to control foreign exchange issues, the Australian government ordered cuts to import licences. It is an illustration of the high esteem in which his very name and his products was held in Australia, that an article in the *Sydney Morning Herald* dated 30 July 1952, broke some dire news to the public.

'WOMEN RUSH 'LAST' HATS OF OTTO LUCAS, 'THE MAN WITH THE MAGIC TOUCH.'

'The last shipment of hats bearing the famous London label of Otto Lucas is expected to cause a stampede among many women for whom THAT import cut will be the one that hurts the most. The collection which was opened in the Elizabeth Street Store of David Jones this week, did nothing to make the parting easier to bear.'

The article went on to trumpet the perfection of Otto Lucas hats – line; craftsmanship; style – time after time, they had the lot.

'When it comes to making hats, he is the finished artist. Even his competitors say of him: "we make good hats – but there is only one Otto Lucas!" His hats turn plain women into elegant women; pretty women into beautiful women; uninteresting women into the kind of women men turnaround and stare after.'

February 1953 saw a coup for Otto Lucas Limited when it was engaged to supply all the hats to grace the heads of models at the first collective British Fashion Industry Show held in the dazzling Suvretta House Hotel in St. Moritz, Switzerland. In a similar vein, Princess Alexandra opened the fifth London Fashion Week in May 1961, attracting trade buyers from thirty-two countries across the globe and – not for the first time – Otto's company provided all the hats.

The degree to which Otto Lucas built a successful business can be deduced from the above, but so, too, can his private lifestyle be judged from other press items. For example, *The Times* (London), 18 September 1957, carried an advertisement for 'a chauffeur/valet required for a gentleman's household in Belgravia. Live in; Rolls [Royce] experience an advantage. Good position for a single man.' In April 1968, *The Times* carried another advertisement for 'a cook/housekeeper and a butler/valet. Must be competent to run a good bachelor household in London SW1. Excellent position for a couple with first class references. Owners away most weekends. Man must be able to drive.'

A particularly revealing article in the *Newcastle Evening Chronicle* of 11 May 1961 gives a rare first-hand glimpse into Otto's private life by journalist Barbara Anne Taylor, who visited Otto in his sumptuous Belgravia home.

Ms Taylor interviewed Otto, an inveterate, heavy smoker, in the panelled dining room, decorated in ice-blue and white and, like the whole house, furnished with French and Italian 'period' furniture, including several white, marble-topped Louis-Quinze tables. The floor was tiled in pale grey and black and a huge gold and crystal chandelier dominated the ceiling above the dining table. All the house windows were draped with white brocade and net and Ms Taylor considered the overall effect was not over-ornate but rather charming – even to her modernist eye. He shares his cool and elegant home with two dogs: Olga and Lola – one, a disturbingly tall, rare poodle and the other, a sad-eyed Dalmation. Though Otto has established himself as a forward-looking trend-setter – who 'has hatted heads from Tooting to Tokyo and most points between' – he lives in a home almost entirely furnished in 'period' furniture. He explained:

> 'Living in a period home has nothing to do with being old-fashioned or reactionary. It is simply that I like to live with beautiful things and I find no grace, charm, beauty or originality in modern interior design.'

His favourite pieces date from the sixteenth, seventeenth and early nineteenth centuries. He said he abhorred fashion designers who hark back to the 1920s in their designs and it made him quite angry to see it still practised in 1961 by many of the Paris fashion houses.

Gaining a reputation as an 'anglophile,' Otto pursued his desire to become a British citizen; applying again in April 1951 (notice in *The Times*; 28 April 1951), before finally achieving his ambition in 1961. The National Archives in Kew has a copy of his naturalisation certificate, BNA66363, which was issued on 25 April 1961. Interestingly, this 1951 notice gave Otto's address as: Chesham House, Lyall Street, London SW1, so it seems he had moved once again, this time to an apartment in another prestigious location in Belgravia.

In tailor-made suits and having a love of opera, poetry, art and champagne, it is perhaps part of his desire to live the life of an English gentleman, that drove him to acquire Hush Heath (or Husheath) Manor, in 1958, a secluded Tudor country house near Goudhurst in Kent. He was still living there in 1971, since the *Kent & Sussex Courier* of 3 September carried an advertisement for a gardener at the Manor. Offering the post at £20 per week (≈ £243 in 2024), it came with a three-bedroom bungalow and a staff car provided for the successful applicant.

Further evidence of Otto's occupation of Hush Heath Manor, came with the startling revelation that the Manor was burgled on the 25 January 1971. Such was his fame, that the crime was reported even by the *Aberdeen Press & Journal*: 'Thieves yesterday grabbed silverware worth £6,400 (≈ £78,000 in 2024), from Goudhurst, the 16th-century manor house owned by Mr Otto Lucas, the internationally-famous hat designer.' Otto was, at this time and according to press reports following his death later that year, resident at 5 Montpelier Place in the Knightsbridge/Belgravia district of London SW7 – yet another prestigious location, where he had been living 'for a number of years.'

It was Saturday, 2 October 1971 and as a new day dawned, Otto was up early in order to be driven by his chauffeur to London airport, Heathrow, in ample time to check in for the departure of British European Airways (BEA) Flight 706 to Salzburg. He was due to be away for a week on business in Switzerland, France and Austria, while hoping also to devote some time to himself in Salzburg to indulge his love of opera and to take a look at some properties that might make a summer retreat.

At 09.34 GMT, Vickers Vanguard 951 turbo-prop airliner, registration G-APEC, took off from runway 28-Left, with eight crew and fifty-five passengers on board (see Appendix 6). It climbed as normal using a *Dover One* Standard Instrument Departure (SID) procedure that routed it via a radio navigation non-directional beacon (NDB) at Epsom and other radio navigation aids called Biggin Hill and Detling VHF omnidirectional radio ranges (VOR). At 09.54 the airliner passed over Dover VOR at 14,200 feet, still climbing on Airway 'Green-One' towards mid-Channel, where radio communication with the ground was handed over from London to Brussels Air Traffic Control (ATC) service at 10.01. Now crossing the Belgian coast, the Vanguard levelled off at its allocated cruising altitude of 19,000 feet over Wulpen VOR at 10.04. The weather was bright sunlight; no turbulence reported and the aeroplane was flying with the autopilot engaged, at an indicated airspeed of 250 knots on its correct course of 110 degrees, with a slight tail-wind of 22 knots.

Five minutes later, at 10.09+46-seconds, with no prior warning, one of the pilots transmitted the chilling message: 'we are going down, 706, we are going down.' This was immediately followed by a 'Mayday' (distress) call, repeated several times by two voices. During the next 54 seconds there were several indistinct messages from the Vanguard, that included: 'we are going down vertically' and 'out of control.' Increasingly-loud background noises ceased abruptly at 10.10+40-seconds, equating to the approximate time of the aeroplane's impact with the ground.

The airliner crashed 30 degrees over the vertical into a grass field at Aarsele in the West Flanders region of Belgium, burying itself in a crater 20 feet deep and being severely burnt when the 25,000lbs of fuel it was carrying, ignited. There were no survivors and one person in a passing car was slightly injured when the vehicle was hit by flying debris.

Significantly, the outer two-thirds of both horizontal stabilisers (tailplane) and the left elevator, together with the whole of the right elevator were not found in or near the crater. These components were discovered several kilometres from the main wreckage and investigators calculated that the tailplanes and elevators had separated from the aeroplane at not less than 18,000 feet altitude. The loss of the aerodynamic downloads normally provided by the tailplanes (in any aeroplane) caused the aeroplane to pitch rapidly nose-down in a steep dive, from which there was no prospect of recovery.

Despite press speculation that the aeroplane may have been blown up by an IRA bomb, the underlying cause, however, was due to the lower part of the rear pressure bulkhead having being affected by liquids that caused severe corrosion and cracks over an unknown period of time. This resulted in the pure chance that the bulkhead ruptured during this flight and not during any one of possibly twelve earlier flights. Inspection work on this aircraft had been correctly carried out according to the maintenance schedules, but the joint British and Belgian air accident investigation team considered that procedures for detecting corrosion in the area of the bulkhead where the fracture occurred, were inadequate. When the corrosion-induced crack exceeded the 'critical crack length,' the rear pressure bulkhead ruptured. Explosive de-compression at that position blew out the bulkhead located at the rear of the passenger cabin. This explosive release of pressurised air from the cabin, caused an abnormal increase in pressure inside the normally un-pressurised tail cone, which the cone and tailplane components running through it were not designed to withstand. Sudden high-pressure air through the tailplane structures blew off their skinning; weakening and distorting them so much that they broke away in flight. The loss of the tailplanes instantly destroyed the aerodynamic stability of the aeroplane, causing it to fall into a dive. This was not a survivable accident.

Similar corrosion was found in other Vanguards and the inspection and maintenance procedures were modified accordingly. Kits to strengthen the bulkheads were ordered and while they were being manufactured and rolled out, all Vanguards were restricted to fly at no higher than 10,000 feet, in order to keep the cabin pressure differential down to 3½ lbs a square inch (psi) instead of the usual 6½ psi; Flight 706 was operating at around 5¾ psi, when the accident occurred.

There was nothing the flight-deck crew could have done and the investigation absolved them entirely. Captain Edward Probert, in command, was only on that flight because he had volunteered to stand in for a colleague, while the supernumerary fourth pilot was Captain Guy Partridge, on board to gain experience of that particular route to Salzburg.

Among the passengers, two whole families including children were wiped out and several couples, too. There is a communal memorial in Aarsele cemetery bearing the names of thirty-seven of the sixty-three victims who were buried in a communal plot, but Otto Lucas is not among them. His remains were among those returned to England, where a service was held for him at the Golders Green crematorium.

Otto was travelling alone and his death represented the end of the road for his business. Otto Lucas Ltd., in Bond Street and Otto Lucas (Foreign) Ltd., in West Smithfield, were both liquidated and wound up voluntarily in June 1972. The Otto Lucas (Junior) company, which offered a more-modestly priced product range, carried on until October 1979 when it, too, was liquidated. Rolf Andresen does not appear to have been involved directly with the business operations. He was, however, a beneficiary of Otto's will and inherited his property, including 5 Montpelier Place and Hush Heath Manor, together with a sum of £150,000 after debts and tax had been paid (≈ £1,700,000 in 2024). Hush Heath Manor was sold in 1987, but Rolf was still resident in Montpelier Place in 2017.

Chapter 18

Graham Hill

29 November 1975

After defying the 'Grim Reaper' on the world's motor racing circuits for over twenty years, Graham Hill OBE, with his debonaire manner and quick wit – the epitome of British Formula One Grand Prix drivers in the 1960s and 1970s era – lost his life in an air crash shortly after his retirement from competitive driving. He was twice crowned World Formula One Champion.

Norman Graham Hill was born in Hampstead, London on 15 February 1929, the son of a stockbroker. Educated at Orange Hill Boys Grammar, Hendon until aged 13, he transferred to Hendon Technical College and after leaving at the age of 16, although at first owning a Velocette motor-cycle, he did not obtain a licence to drive a motor-car until 1953, at the age of 24. Out riding his motor-bike one foggy night, he ran into the back of an un-lit stationary vehicle, broke his thigh and was left, after three months in hospital, with a slightly-shortened left leg and a back problem that recurred from time to time. After Hendon Technical College, Hill joined the Smiths Instruments Company in London on a five-year engineering apprenticeship, until conscripted into the Royal Navy in 1947. While in the Navy, Graham took up the sport of rowing with a club in Southsea. Having completed his national service, now back working at Smiths he became an enthusiastic oarsman, first with the Auriol Rowing Club in Hammersmith, where he met his future wife, Elizabeth (Bette) Shubrook – herself a very good rower – and then joining the London Rowing Club in 1952. During his future nomadic lifestyle, Bette was extremely supportive and they raised a family of two daughters: Brigitte and Samantha and one son: Damon. This period in his sporting life accounts for another one of Graham's most-recognisable personal flourishes: his use, throughout his racing career, of the London Rowing Club colours, seen painted as eight white, vertical oar-blades spaced round a dark blue racing helmet.

In 1953, responding to an advertisement by Universal Motor Racing Club at Brands Hatch circuit, offering trial drives at five shillings a lap, he blew £1 to buy four laps in a Cooper 500 Formula 3 racing car and from that moment, committed all his energies and indeed his considerable persuasive charm, to becoming a competitive racing driver. He persuaded Universal to employ him as a part-time mechanic; edging himself into a position as an instructor, but he had to move to another job at the circuit with Don Parker, the 1952 and 1953 500cc Formula Three champion, before he finally had the chance to squeeze himself into the cockpit of a car for a few competitive races. After one of these early races – where his results reflected his lack of experience – he met a fellow driver called Colin Chapman. Besides driving, Chapman was trying to develop his own racing car which he called a Lotus. The two men got on well enough for Graham to talk his way into working part-time as a mechanic for the Lotus operation. Inevitably, Graham tried to persuade Chapman to let him drive a Lotus competitively, but Chapman was reluctant to entrust his new car to Hill's inexperienced hands. However, choosing to pursue his dream of a career in motor racing rather than working for Smiths, Hill joined Colin Chapman full-time and from 1956, was even allowed to race sports cars occasionally for the Lotus stable; his first competitive race coming on 29 April 1956 with a win in a Lotus at Brand's Hatch.

Colin Chapman recognised Graham's potential and in 1958, Hill was offered a full-time driver role; making his first Formula One (F1) start, in a Lotus-12 Climax car at the Monaco Grand Prix (GP). He competed in nine races that year but was forced to retire in seven of them.

In 1960 he was invited to join the Owen Racing Organisation's British Racing Motors (BRM) team based in Bourne, Lincolnshire. BRM were developing their new P-48 car and felt that Hill's experience as a mechanic as well as a driver was ideal to help this project along. He had a much better season, qualifying for a front row start in Argentina; seventh at Monaco and had a podium finish with his third place at Zandvoort in Holland. All his other starts ended in retirements but he gained four points in the world driver's championship.

1962 saw the operational debut of BRM's long-awaited V-8 powered P-57 car and another milestone was passed when Graham Hill won his first F1 race at the Dutch GP in Zaandvoort.

On Sunday, 5 August 1962, Hill and Jim Clark jostled for the world title at the Nurburgring in Germany, with Graham starting the fifteen-lap race (it was a 17-mile circuit) ahead in the points table. After 2½ hours and fifteen gruelling and exhausting laps in very wet conditions, Graham crossed the line in first place, with just 4.4 seconds covering the first three cars. Jim Clark took fourth place and only sixteen of the twenty-six starters finished the race.

Clark took the USA GP at Watkins Glen and the outcome of the world championship of 1962 went down to the last race of the season in South Africa. Jim Clark led the way but suffered mechanical failure and he retired. Graham Hill took the lead and held it to the chequered flag. At the age of 33 and having been an F1 driver for four years, he had achieved his dream. He was the Formula 1 world champion.

The year 1963 brought technical problems for BRM and race reliability fell away, although Graham savoured the first of what would become a record five Monaco GP wins.

BRM's car was on song in 1964 and the championship turned into a dramatic Hill versus Clark duel once again. Even though Graham Hill accumulated more points during the season (forty-one), the F1 rules for this year counted points from the six best results (thirty-nine), which meant that John Surtees (forty; Ferrari) pipped him to the title by just one point. It was not all gloom, though. Hill finished second in the Le Mans 24-hour race with his Swedish co-driver, Joakim Bonnier; also winning that year's Reims 12-hour and the Paris 1,000-kilometre races, all in Ferrari cars. Graham Hill had now firmly established himself as a force to be reckoned with on the world motor-sport stage.

The next year, 1965, Graham Hill won the Monaco GP again but, even though he finished every race except one and took second place in the driver's championship for the third successive year, Jim Clark was quite unstoppable in his own quest and clinched the world title.

The 1966 F1 season proved a fallow one for Graham when he failed to gain a win during the whole year. There was, however, some compensation in terms of both prestige and finance, when Hill was offered a late invitation to drive a Lola T-90 Ford in the Indianapolis 500 – and won. Starting from fifteenth on the grid, Hill knew he was in a dangerous, high-speed race of attrition; a race with a notorious reputation for crashes and fatalities, but he dodged all the carnage to become the first 'rookie' winner since 1927, at an average speed of 144mph.

By this date, Graham had learned to fly and gained his private pilot licence (PPL) in England and used some of his winnings from the Indy 500 to buy a light aircraft with a view to using it to travel to and from race circuits. The winner's purse in 1966 was $156,000 (≈ $1.5 million in 2024) and it has been suggested that Graham's share amounted to £55,000 (≈ £862,000 in 2024).

Graham Hill changed stables again in 1967 by re-joining Colin Chapman at Lotus, where he was now in the company of his former rival, Jim Clark. In his eleven starts, Hill was dogged by retirements in eight due to mechanical problems and only managed two second placings at Monaco and Watkins Glen.

The 1968, twelve-race season got off to a cracking start, with the Lotus-49 cars of Clark and Hill clinching first and second places at the Kyalami circuit in South Africa on 1 January. Disaster struck the team, though, on 7 April, when twice-world-champion Jim Clark was killed during a Formula 2 race at Hockenheim. The motor racing world – even though accustomed to all-too regular fatalities in those days – was stunned by the loss of the revered Clark. The whole Lotus team, including Graham Hill, took it very badly. Hill, now left as the lead driver, swallowed his grief and pulled the team's morale up by its boot-laces. Spurred on by Graham, the team went into the second race in Spain with great fortitude and determination and he, too, gave it his all – would he ever do less? – to take first place. He did the same again in Monaco; took second places in Germany and the US and even with four retirements, he topped a magnificent season by winning the final GP in Mexico to clinch his second world title with forty-eight points. By the force of his courageous personality, he had led his team from the front and brought the title back to Lotus. Graham Hill was at the pinnacle of his profession, the public adored him and the country honoured him when, in her Birthday Honours list of 31 May 1968, HM the Queen appointed him an Officer of the Order of the British Empire (OBE) for his services to motor racing.

Graham Hill even tried the limelight provided by the movie industry. He first took to the silver screen in 1966, when he played racing driver 'Bob Turner' in the feature film *Grand Prix*, starring James Garner in the role of a champion F1 driver. There were also 'bit parts' for his racing buddies: Jim Clark and Jack Brabham in the same film. Graham had another dabble in the movie world when in 1974, he played the part of a helicopter pilot in the Rank Organisation's film adaption of an Alistair McLean best-seller: *Caravan to Vaccares*. There were other glimpses of Graham Hill the actor in, for example, *Threads* (BBC Nine Network;1984) a sci-fi drama documentary about the effects of a nuclear attack on Sheffield, in which he played 'first soldier'. *Bon viveur* and *raconteur* – he was considered one of the best speech-makers in London – and his presence was guaranteed to enliven any function. Graham's dashing character, good looks and witty repartee, brought him much exposure on television, too, such as being a panel member for at least sixteen programmes of the popular BBC TV show *Call My Bluff*. There was also a one-hour biopic about his racing career narrated by movie legend, Paul Newman, which aired in 1974 and he participated in many radio and TV chat-shows about his life and career.

In some accounts of his racing life, it is suggested that Graham Hill was not what was called a 'natural driver' and that he 'had to work hard' at the task. But such a view is considered unfounded, for who dares to define a 'natural' driver and what is meant by that term anyway? In whatever terms Graham Hill's driving style is described, one cannot get away from the fact that he drove against contemporaries from all over the world – many of whom have been also been described as 'naturals' – and won two world championship driver titles. He raced in Formula 1, the Le Mans 24-Hour; the Indianapolis 500; the BRDC International Trophy and saloon car competitions all over the world and won in all of them. Indeed, Graham Hill is the only driver to achieve the so-called Triple Crown of motor-sport – the definition of which has also been open to interpretation over the years. There are two opinions among the racing fraternity as to what this accolade should include. With Hill's achievements shown in brackets, the first opinion considers that a driver must win: the Monaco F1 GP (1963-65,1968 & 1969); the Le Mans 24-hour race (1972) and the Indianapolis 500 (1966). The second opinion considers that a driver must win the F1 world driver championship (1962 & 1968); the Le Mans 24-hour and the Indianapolis 500. Graham Hill did it both ways! Perhaps it should simply be said of Graham Hill that he had his own style of talent – and it worked for him.

The new season brought problems for Hill and his team-mate Jochen Rindt, with the rear-aerofoil attachments on their cars. During the Spanish GP at Barcelona in 1969, the rear unit fitted to both Lotus cars collapsed, causing bad accidents from which both drivers were lucky to emerge with only minor cuts and bruises. In November, Hill was less fortunate. By lap eighty-eight of the 108 lap US GP at Watkins Glen, with the tyres wearing, Graham found the Lotus difficult to handle and spun off on a patch of oil, stalling the engine. Unlocking the safety harness, he climbed out of the car and pushed it to re-start the engine. Climbing back in, he could not re-buckle the harness without assistance, so he left it undone and continued the race. Passing the pits, he signalled his intention to come in for a tyre change, but shortly afterwards a tyre burst at speed; the car cartwheeled and crashed. Graham was thrown from the car and suffered a broken right knee and a dislocated left knee. Hospitalised in the USA for three weeks, he remained very upbeat and was even interviewed for a British radio broadcast, which greatly enhanced his standing with the British public.

Although he competed in the 1970 season for Rob Walker's independent Lotus team, Graham was thought to be far from fully fit, but in his own inimitable way, he pressed on. In 1971 he moved to the Brabham team, staying with it for the 1972 season as well, but his results were poor compared to those before the accident.

Despite suggestions that he was past his prime, Graham had the satisfaction of handling wet conditions at Le Mans and with his co-driver Henri Jacques Pescarolo (Fr), drove a French Matra-Simca MS670 to victory in the 24-hour race of 10/11 June 1972. A disgruntled Pescarolo felt the 43-years-old Hill was not up for taking the risks involved in this dangerous and demanding race particularly if it was to rain or be foggy, but he could not have been more wrong. There was nothing wrong with Graham's will to win and he brought experience and the desire to grasp this – probably last – opportunity to write himself into the record books as the only man to win that elusive Triple Crown. By the time the light began to fade that evening, the lead cars were all Matras. The Hill/Pescarolo pair took the lead and never looked back. With the car being driven brilliantly by both men, their victory was assured and with it came Graham's unique Triple Crown title.

It was around this time that he felt he was ready to set up his own racing enterprise so, in 1973, with sponsorship from Imperial Tobacco Company he formed his own F1 team: Embassy-Hill. It was ready to compete in April 1973, with Graham as the lead driver but his results were off the pace during that fifteen-race season and also for the next, when two cars were entered.

When Graham did not qualify for the Monaco GP of 1975, he decided it was time to retire from competitive driving and to concentrate on managing his team, which had now acquired two new, up-and-coming drivers: Tony Brise and Alan Jones – the latter going on to win the world title in 1980 with the Williams team. With these promising drivers and a new-design car, the Cosworth-engine Hill GH1 designed by Andrew Smallman and based on a Lola T370, things looked bright for Team Hill. However, although entered in twelve F1 GPs in 1975, the GH1 did not deliver the success it was hoped for, but Smallman was already working on a new design, the GH2.

The GH2 was shipped out to the *Paul Ricard* circuit in the south of France, for testing. On 28 November 1975, Graham Hill flew himself and the key members of his team, in his Piper PA-23-250D Aztec, N6645Y, to Le Castellet airport, near the circuit. Tony Brise telexed good news to the team base: during testing, the car performed very well; as if all its ills were solved by the latest modifications. Since there was now no reason to prolong the trip, Hill and the team prepared to fly back to London on 29 November.

According to the narrative of the UK Dept of Trade, Accident Investigation Branch, Aircraft Accident Report 14/76, departing at 15.30, Hill and his passengers flew first to Marseille-Marignane airport, so that he could obtain fuel, weather reports and file a plan for an IFR flight to Elstree airfield, London, with Luton airport as the alternative destination. Graham's passengers that day were: team manager: Ray Brimble; car designer: Andrew Smallman; team driver: Tony Brise and team mechanics: Terry Richards and Tony Alcock.

Take off from Marseille was at 5.47 pm, for an expected trip of four hours duration. All went well and approaching the coast, Graham made radio contact with London air traffic control centre at 8.45 pm. Reporting over Dover at 9.00 pm, he was cleared to Elstree via Lambourne VOR (Very high frequency [VHF] Omni-directional Range) radio navigation beacon and was advised that Elstree had 2 kilometres visibility and the cloud-base was 300 feet above ground level (agl). When asked what his intention was after Lambourne, he said he would take a look at Elstree and Luton remained his alternative.

Approaching Lambourne VOR, Graham was advised that visibility at Elstree was now 1,000 metres. Handed over to London Heathrow approach, at 21.20, he was instructed to descend to 4,000 feet and that visibility was now 800 metres at Elstree. Given various course changes by Heathrow approach, to bring him onto a heading of 280 degrees for Elstree, at 21.22 he was instructed to descend to 1,500 feet. At 21.27, Graham reported at 1,500 feet and was advised he was 4 nautical miles from Elstree. After Graham asked for further descent clearance, the controller advised him that further descent was at his own discretion. At 21.28, the controller informed him that Heathrow's QFE – international code for the barometric setting for aircraft altimeter – was 999 millibars (mb) and there were 3 nautical miles to run to Elstree. This transmission was not acknowledged and shortly afterwards, when the radar plot showed the aeroplane at 2½ nautical miles from Elstree, the plot disappeared from radar screens.

After his last exchange with Heathrow approach, Graham called Elstree air traffic control and was heard to report '4-5-Yankee, finals', to which the Elstree controller, recognising the pilot's voice, responded '4-5-Yankee, roger, Foxtrot Echo 990 [QFE: 990mb], check three greens, clear to land.' This transmission brought the response of '45' followed by a click – and silence. There was no further contact between the aeroplane, Elstree or Heathrow approach.

It was later discovered that, at about 21.30, the sound of a low-flying aeroplane was heard by three people on the ground in the Barnet area and they caught a glimpse of a flashing light through the fog, travelling east to west in the direction of Arkley. Another person, located 650 metres east of the accident site, heard a low-flying aeroplane, followed by what seemed to be the sound of an explosion in the direction of Arkley golf course. The weather at this location was dark, with visibility of 50-100 metres in thick fog. When Heathrow approach established with Elstree that 4-5-Yankee had not landed, emergency procedures were initiated. By 22.00, police confirmed the worst; that the Aztec had crashed on Arkley golf course. Three nautical miles east of Elstree airfield threshold and one nautical mile south of runway-27's extended centre-line; wings level and descending, it had struck trees, first at a height of 458 feet above mean sea level (amsl), then hitting more trees before rolling to the right and impacting the ground. It was completely destroyed and all six occupants died instantly.

The UK Department of Trade, Accident Investigation Branch aircraft accident report number 14/76, concluded that it was not possible to establish the exact cause of the accident, but the possibility that the pilot underestimated his range from the airfield and in consequence descended prematurely in conditions of thick fog, cannot be excluded.

Graham was laid to rest in the graveyard of St Botolph's in Shenleybury, Herts now, since the church was made redundant, a private residence but retaining access to the cemetery. A legend of world motor sport was lost and his racing team all but wiped out at a stroke. Team Hill never recovered from the blow of losing its charismatic leader and its key members. Its assets were acquired by the Canadian Walter Wolf racing team but the GH2 car never raced competitively, so its F1 potential was not put to the ultimate test. It did, however, remain intact, at one time being displayed in the British Motoring Museum; even being driven occasionally at classic car events. In 2004 it was in the hands of Klaus Fiedler and still being driven but was last heard of in 2015, when it was sold to an un-named private collector in a specialist motoring auction – ironically held in premises located on Graham Hill Way in Bourne, Lincolnshire!

Graham Hill's Grand Prix career spanned eighteen seasons. He competed in 176 GPs – and ten years would elapse before another driver, Jacques Lafitte, passed his total – with 14 GP wins, 36 podium places (finishing in the first three) and two F1 Drivers World Champion titles.

Chapter 19

Francis Gary Powers

1 August 1977

It was 08.15 on 10 February 1962; a cold and dismal morning in East Berlin. Several cars drew to a halt at the eastern end of the Glienicke Bridge. Wearing a heavy overcoat and a Russian-style fur hat; carrying a small case and a rolled-up rug, Francis Gary Powers got out of one of the cars. The Soviet army colonel who had accompanied him from Vladimirsky prison, got out, shook his hand and they walked towards the bridge. It was closed to traffic that morning so that an unusual ritual could be enacted. Francis Powers – known to the world as 'the U-2 spy pilot' – was about to taste freedom for the first time in nearly two years.

Francis Gary Powers, was born on 18 July 1929 in Burdine, Kentucky, USA, a coal-mining community on the Kentucky-Virginia border. His mother was Ida and his father, Oliver, had worked down the pits virtually all his life. Francis was the only son in a family of six children and it was his father's hope that his boy would never work in the mines, but would become a doctor instead.

In 1946, Francis started a medical course at a college in Milligan, Tennessee, but his heart was not in it. Taking the plunge, he dropped his medical studies but continued with biology and chemistry and he decided that when he graduated, he would join the United States Air Force (USAF). He wanted to fly.

In October 1950, aged 21, Francis Powers BSc – now preferring the name: Frank – signed up for a four-year engagement with the USAF as a photographic laboratory technician. Keen to transfer to the flying branch, he applied for Aviation Cadet flying training; was accepted and began his air and ground training with the 3300th Training Squadron at Greenville AFB, Mississippi. In December 1952, Frank graduated as a fully-fledged pilot with the rank of Second Lieutenant in the Air Force Reserve, flying the Republic F-84 fighter jet.

From 5 to 15 October 1953, having had his security status duly raised, Frank attended a Special Weapons Delivery Course at the Air Force Special Weapons Centre at Sandia AFB, near Albuquerque, New Mexico. Here he received instruction on how to deliver a tactical nuclear bomb from an F-84G.

In July 1954 Frank was promoted First Lieutenant and on 2 April 1955 he married Barbara Gay Moore. It would turn out to be a stormy relationship that was doomed to failure. Frank's four-year enlistment term was due to end in December

1955 and he was interested in finding a flying job in the civilian airline industry but, he found his age (almost 27) went against him. He was very keen to keep flying so, giving up on the idea of civilian flying, he decided to continue in the USAF Reserve and re-enlisted on 18 December 1955, this time for an indefinite term.

It was in January 1956 that Francis Gary Power's career path now took an unexpected and monumental turn.

Together with several other pilots on the squadron, Frank found his name on a list ordering him to the adjutant's office in the Wing Headquarters building. The sense of mystery deepened when the adjutant announced that, on the strength of their individual Air Force qualifications (Frank was graded as an 'above average' pilot) and Top-Secret security clearance, some people – unspecified – were interested in talking to them about a new flying job – also unspecified. His curiosity and innate sense of adventure aroused, Frank said he was keen and so was permitted to go off-base, alone, to meet a certain Mr William Collins in a local motel.

Frank met Collins and was informed that the job – still unspecified – was 'risky; patriotic; better paid; overseas and un-accompanied.' Powers would be working for the CIA on an eighteen-month civilian contract. He would have to leave the Air Force but would, at the end of his CIA contract term, have the right to return to the service with no loss of rank or benefits. Collins outlined the job and what was expected of him. Most startling was that he would be trained to fly a new aeroplane that could fly higher than any other and his main task would be flying photo-reconnaissance and electronic-monitoring missions over Soviet territory – among other places – consistent with the US government intelligence-gathering programme known as Operation *Overflight*. Travelling under the alias of 'Francis G. Palmer,' he was to report to a Washington DC hotel where he would begin the process of joining the Agency. Francis Gary Powers was about to become a spy.

Frank's introduction to and training on, the Lockheed U-2 was carried out at a remote, CIA-run, secret airfield at Green Lake in the Nevada desert and lasted until August 1956.

Designed and built by Lockheed for an ultra-high-altitude, all-weather reconnaissance role, the specification, performance and characteristics of the single jet-engine U-2 aeroplane are widely-documented elsewhere and will not be repeated at length here. With a range of around 3,000 nautical miles (nm), it was built for lightness rather than robustness or longevity and as a result, its cruising altitude was spectacular. Current US official thinking was that no Soviet aeroplane or missile could get anywhere near it. Designed to operate at altitudes up to 74,000 feet, the U-2 cockpit was pressurised and pilots wore a partial pressure suit with a fully enclosed helmet. Prior to missions, it was vital to control liquid intake since the suit could not be un-zipped for the call of nature. It was also necessary to start a process of 'pre-breathing' pure oxygen for two hours, in order to prevent onset of a physiological condition known as 'the bends.' Navigation on overflight missions was manual map-reading, assisted occasionally by on-board radio-compass fixes on Soviet civil radio-station broadcasts, since – for obvious reasons – no external radio navigation aids could be used over hostile territory. Perhaps the most interesting switches in the cockpit were those operating the self-destruct explosive charge that was supposed to demolish the bay containing the camera and electronics apparatus. Testing had suggested the optimum delay for the pilot to 'get the hell out of it,' was 70 seconds after the switch was activated.

Francis was allowed two-weeks leave before departing for Turkey, to join what was known to the outside world as a weather-reconnaissance unit designated 2nd Weather Observational Squadron (Provisional), based at Adana AFB, later re-named Incirlik. This unit of five aeroplanes and seven pilots, also referred to as Detachment 'B,' functioned as a combined USAF/CIA operation, with the former providing logistics and the latter the planning of operations. By now, although President Eisenhower had 'bought-in' to the whole U-2 concept of intelligence gathering, he frequently wavered over CIA requests for deep penetrations, being torn between his desire to maintain a 'correct and moral' position on the world stage and his wish to know what the hell the Soviet Union was up to.[1]

Francis Powers made his own first operational intelligence-gathering flight in a U-2 on 27 September 1956, which was a sortie lasting 6h 10m – believed to have been a reconnaissance watching the build-up to the Anglo-French-Israeli invasion of Egypt and the Suez Canal area. His flying skill was highly regarded by his superiors, to the extent that he was the pilot selected to carry out Detachment B's first overflight of Soviet territory on 20 November 1956. A further example of Francis' involvement in U-2 operations is illustrated by his flying logbook entries for August 1957. He flew an 8h 35m overflight of Soviet territory on the 5th, then on 16th he logged 6h 40m as co-pilot in a Lockheed C-124 Globemaster II, which carried special fuel for the U-2 from Adana to Peshawar.

Between 26 June 1956 and 1 May 1960, the whole U-2 force – based in West Germany, Turkey and Japan – logged fifty-one operational flights in total. Of these, twenty-six flights made incursions into Soviet territory and it is believed that Francis Powers may have flown twelve of these latter missions, out of his own twenty-seven operational flights.

In 1957 the USAF had commenced flying operations from two forward airbases at Lahore and Peshawar in Pakistan. It was from Peshawar that the fateful flight, code name Operation *Grand Slam*, was flown on 1 May 1960 and projected Francis Gary Powers onto the world stage.

To date, Francis had accumulated almost 600 flying hours in the U-2 and as the unit's most experienced pilot, he was the 'go-to guy' for this special mission, the 26th overflight of Soviet territory. This one was special, too, because it was the first mission to be routed one-way over hostile territory. It would go in from Pakistan and out in Norway. More than nine hours and 3,800 nautical miles of bum-numbing tension. To get the extra range needed, two 'slipper-tanks' (a normal addition) were fitted, one on each wing, each with a capacity for 100 gallons of fuel.

The story of this flight actually begins on 28 April. Individual U-2s were identified by the Lockheed, three-digit, production number known as the 'Article number' as well as their USAF serial number. Article 358, USAF serial 56-6691 was ferried from Adana to Peshawar on 28 April. On 29 April, the mission was delayed for a day because Bob Ericson flew 6691 back to Adana, where it was taken out of service for routine maintenance. A replacement U-2C, Article 360, 56-6693, was ferried out to Peshawar but, on 30 April the mission was scrubbed due to poor weather over the Soviet Union. The forecast looked good for Sunday, 1 May.

Taking off from Peshawar at 07.00, the U-2 headed towards Afghanistan, rapidly gaining height towards 70,000 feet over the Hindu Cush, before crossing the Soviet border near Dushanbe in Tajikistan. Heading north he flew over Stalinabad towards Tyuratam, in Kazakhstan. This was a key target, since its Baikonur Cosmodrome was a major missile testing site for ICBMs and space research. Visible in the distance, the Aral Sea provided Francis with a reference point for a course correction and he adjusted his track northwards for a long run towards Chelyabinsk, east of the Urals; a plutonium production facility for nuclear weapons. Sverdlovsk (Yekaterinburg) was the next target, a major industrial area 200 miles further north and site of the main uranium enrichment facility for the Soviet nuclear weapons programme. His route beyond this point was to Kirov/Yurya; Plesetsk missile test centre; Severodvinsk submarine shipyard; Murmansk naval base to Bodo (Norway).

It was about 1,300 miles into the mission and at 70,500 feet altitude, that the auto-pilot failed; requiring Francis to fly the aeroplane manually. Forty miles south-east of Sverdlovsk was a turning point on the course, that would take him across the target area on its southern edge, lining him up on a north-west track to the next target: an ICBM base under construction at Yurya. At each turning point, he noted various cockpit instrument data and that, coupled with switching on the cameras, occupied his attention.

A minute or so after rolling out onto the new heading, he sensed, what felt like an explosion – but not, he thought, from inside the aircraft. A wing dipped and he corrected it. Then the nose dropped – and kept going down. He could not correct it with the control column; suggesting to him that perhaps the tail section had broken off. Diving ever faster, Francis thought the wings parted company at this point. Despite his seat harness, he was thrown violently around the cockpit as the U-2 fell into a fast, nose-up spin. 'G'-forces made it impossible to take up the correct body-position for an ejection, so, he decided to jettison the canopy and go over the side – checking this was possible before he committed to throwing the destruct switches.

FRANCIS GARY POWERS: 1 AUGUST 1977

With the altimeter unwinding past 34,000 feet; no time to think; everything happening fast. He was sucked half-out of the cockpit – still tethered by his oxygen tube. Trying to reach back inside to throw the destruct switches, he found the 'g'-force stopped him. It was time to go! His helmet face-plate frosted over and after several lunges, his oxygen pipe finally snapped and he shot away from the plummeting aircraft. A trailing line pulled the ripcord and his canopy blossomed at about 10 – 15,000 feet.

Eight missiles were fired at the U-2, but it was brought down by a near-miss from one of a salvo of three S-75 *Dvina* surface-to-air missiles that exploded near its tail. The resultant shock-wave was enough to cause structural failure of the lightly-built aircraft fuselage. Due to an out-of-date code in its defensive electronics, a Soviet MiG-19 jet fighter was also hit by one of the missiles and brought down with the loss of its pilot, Lieutenant Sergei Safronov.

Soon after landing outside the village of Kosulino, his pistol, knife, survival pack and parachute were confiscated and a crowd, growing to about 200 people, prevented any possibility of escape. Within three hours of landing, he was aboard an airliner bound for Moscow and the Lubyanka prison, where from 3 May, he was interrogated by the KGB every day for sixty-two days. Initially interrogation sessions occurred several times, through the day and night; occupying up to 12 hours in 24 but, as time passed, these sessions grew shorter and more spread out. He remained in the Lubyanka in solitary confinement until 9 July.

The political fallout from the crash was swift and highly embarrassing for the USA and personally for President Eisenhower. His worst nightmare had happened.

Francis Powers, indicted before the Military Division of the Supreme Court of the Soviet Union, appeared in a public 'show' trial held, from 17 to 19 August 1960, in the Hall of Columns, in the House of Unions building in Moscow. It was a grandiose affair that captured the attention of the whole of the world's media. Francis' wife, parents, mother-in-law were permitted to attend and the Russians appointed a lawyer to represent Francis. The verdict of the three judges was inevitable. On 17 August 1960, he was found guilty of espionage and sentenced to ten years in jail: three years to be served in prison and seven as hard-labour in a prison camp. He was lucky; it could have been the death penalty!

Francis was transferred to Vladimirsky prison, 150 miles east of Moscow, where he shared cell-31 with a Latvian prisoner, whom he came to like very much.

Out of the blue, at 19.30 on 7 February 1962, Francis was whisked off to Moscow – without guards – and next morning, the 8th, he was flown to East Berlin. On Saturday 10 February, after twenty-one months of captivity, Francis Powers was exchanged – in the first 'spy-swap' made between the US and USSR – for Rudolf Ivanovitch Abel, a Soviet spy caught in the USA in 1957 and sentenced to 45 years in prison. Abel was later identified as Colonel Vilyam Fischer, a KGB operative, born William Fisher in Newcastle, England in 1903.

FATAL FLIGHTS OF THE RICH AND FAMOUS

The spy-swap took place at the Glienicke bridge in the Potsdam district of Berlin and it was during the final few minutes that CIA security officer, Joe Murphy, did his bit with personal questions. Joe said:

> 'I was selected to identify Powers during the exchange, because I had been with him while he flew U-2s from Turkey. On 10 February 1962, I joined the Chief of the US Mission to West Berlin, Alan Lightner and Jim Donovan, the lawyer who had negotiated the exchange, on the Glienicke bridge. [After the exchange] Powers and I went to Rhein-Main AFB and flew back to the USA.'

For Francis, life was never quite the same again. He was famous the world over. Debriefed by and the subject of inquiries by, the CIA, USAF, Lockheed, the Senate and the House; given lie-detector tests; subjected to constant attention by the media and the public, Francis was called to account for his actions before the Senate Armed Services Select Committee, in Washington DC on 6 March. After questioning in the Senate hearing, the following statement was issued:

'On all the information available, it is the conclusion of the board of inquiry which reviewed Mr Powers' case and of the Director of Central Intelligence, who has carefully studied the board's report and has discussed it with the board, that Mr Powers lived up to the terms of his employment and instructions in connection with his mission and in his obligations as an American under the circumstances in which he found himself. It should be noted that competent' aerodynamicists and aeronautical engineers have carefully studied Powers' description of his experience and have concluded on the basis of scientific analysis that a U-2 plane damaged as he described would perform in its descent in about the manner he stated. Accordingly, the amount due Mr Powers under the terms of his contract will be paid to him.'

His back-pay came to $50,000 ($\approx$ $433,000 in 2001).

Despite all of this, findings exonerating Powers were given limited publication and the sensationalized press of the day resulted in a negative public portrayal of Powers. His motives and loyalties were questioned. His divorce in January 1963 from Barbara further stained his reputation – unfairly – in the press, although on 26 October that year Francis re-married, his new wife being Claudia Edwards Downey, a divorcee with one daughter. Their son Francis Gary junior was born on 5 June 1965.

When the fuss died down, Francis did not re-join the USAF. Instead, Clarence 'Kelly' Johnson offered him a pilot job with Lockheed, testing modifications and new equipment on the U-2 programme, at a salary of $35,000 a year. He accepted and joined Lockheed on 3 November 1962, remaining with them until February 1970. In response to his request, on 25 April 1963 the Director of the CIA sent Johnson the following message.

FRANCIS GARY POWERS: 1 AUGUST 1977

'Approval hereby given under various U-2 contracts for Francis Gary Powers to fly U-2 aircraft, except those contracts involving Operation '*Oxcart*'.

Oxcart covered the development of the Lockheed A-12 that evolved into the SR-71 Blackbird reconnaissance aircraft.

Having resisted publishers for many years, Francis finally wrote his memoir (*Operation Overflight*), which was published during 1970. The CIA was not happy about this and allegedly, Johnson was persuaded to make him 'surplus to requirements.' Having left Lockheed, Francis was unemployed between March 1970 and June 1971 but in July – having divorced Barbara and then married Claudia 'Sue' Downey in 1963 – he and his family re-located to Studio City, in the San Fernando district of Los Angeles where he found a flying job as an airborne traffic reporter with a local radio station: KGIL. As Southern California's Big Red Skywatch pilot, Francis regularly flew a Cessna 172, N44615, on traffic watch duties, handling all air traffic control approaches and departures deftly while making his traffic reports; well-known, too, for his closure of: 'this is Gary Powers, KGIL Skywatch, signing off.'

In 1976, aged 46, Francis joined local TV station: KNBC. The station introduced the world's first 'telecopter,' an 'eye in the sky' television camera and broadcast transmission system mounted in a Bell 206B JetRanger helicopter, registration N4TV. On 16 October, Francis began his conversion training in a Bell 47G and after qualifying as a helicopter pilot, made his first telecopter flight on 26 December, followed by his first live broadcast flight on 11 January 1977.

In the days leading up to and including the fateful day of 1 August 1977, Francis had been covering severe wild-fires in the Corona district, with his cameraman George Spears. It was said that he had earlier reported that the helicopter's fuel gauge was inaccurate; that it registered empty when it still had twenty minutes of fuel left in it. He flew it in this condition and made allowances for it. The fault was repaired – 'empty' now really did mean 'empty' – but it is alleged the mechanic did not inform Francis that it had been fixed. On the day of the accident, getting low on fuel over the fires, he set off for his base, passing over several airfields where he could have landed and re-fuelled. It is believed he pressed on to his base at Van Nuys airfield believing he had enough fuel in the tank – but he had not. The helicopter ran dry about a mile short of the airfield. He should have been able to bring the aircraft to the ground safely, by 'auto-rotation,' the process of keeping the now-unpowered rotor blades turning by air-flow in a deliberate descent. Evidence from the wreckage, however, suggested that the rotor-blades were not turning at a speed consistent with auto-rotation but, that the helicopter was in free-fall. The NSTB report noted the probable causes of the accident as: fuel exhaustion; mismanagement of fuel; improper in-flight decisions or planning and improper operation of flight controls. The telecopter, N4TV, crashed on Sepulveda recreation field at Encino around 12.40 pm, killing George Spears and Francis Gary Powers.

FATAL FLIGHTS OF THE RICH AND FAMOUS

Upon his return from captivity, Francis Powers was made the scapegoat unfairly for poor decisions at high-level in the US administration, but bore it stoically. He was a patriot with legitimate gripes. Over time the US government made some amends but these came mainly after his untimely death. In 1965, some years after his U-2 colleagues, he was awarded the CIA's Intelligence Star for Valor. In 2000, his family was presented with the posthumous awards of the Distinguished Flying Cross, Prisoner of War Medal, National Defense Service Medal and the CIA Director's Medal for Extreme Fidelity and Courage. Francis was also posthumously promoted to the rank of Captain – as would have been his right in relation to years served in the USAF/CIA. As late as 2012, a posthumous USAF Silver Star was also presented to the family, with its citation summing up the belated change of stance. 'Mr Powers,' it said: 'was interrogated, harassed and endured unmentionable hardships on a continuous basis by numerous top Soviet Secret Police interrogating teams, resisting all Soviet efforts through cajolery, trickery and threats of death, he exhibited indomitable spirit, exceptional loyalty, and continuous heroic actions.' Francis Gary Powers was buried in Arlington National Cemetery among his nation's other heroes.

Chapter 20

Payne Stewart

25 October 1999

F-16 pilot, Captain Chris Hamilton of the 40th Flight Test Squadron, USAF, was 'dog-fighting' an A-10 Thunderbolt II over the Gulf of Mexico, when he received a message to break off the exercise; rendezvous with a tanker over Eglin Air Force Base (AFB), Florida; take on fuel and then chase down a civilian Learjet that was not responding to radio calls.

At 10.52 Eastern Daylight Time (EDT), vectored by civilian air traffic control to within eight nautical miles of the Learjet, fifty minutes and 400 miles later, he was flying off its wingtip, 46,000 feet over Tennessee. His report was chilling. Despite his radio calls to the biz-jet and closing to within 100 feet, manoeuvring all around it, he could elicit no response whatsoever. The cockpit and fuselage windows seemed to be fogged up inside, or iced over and nothing could be seen of the interior. No-one seemed to be flying the aeroplane!

Hamilton maintained station on the Learjet at a safe distance until, low on fuel, he was directed to land at Scott AFB, Illinois. Two more F-16 fighters, this time from the Oklahoma Air National Guard (ANG) were diverted to the target. They reported the same situation before having to break away to find their in-flight tanker to re-fuel. Now two F-16s from North Dakota ANG arrived and shortly afterwards they were re-joined by the first two fighters. Keeping a safe distance from the Learjet, which was porpoising gently between about 45,000 and 48,000 feet due to the competing forces of wing aerodynamics and the thin air, the bizarre formation flew northwards, until the lead F-16 pilot reported: 'target is descending and he is doing multiple aileron rolls; looks like he is out of control [and] in a severe descent.'

At 13.12:26 EDT on Monday, 25 October 1999, Learjet 35, registration N47BA, carrying the world-renowned American professional golfer Payne Stewart as a passenger, hit the ground near Aberdeen, South Dakota. It had been airborne for 3 hours and 54 minutes, travelling 1,500 miles over eight states at a speed of around 400mph and reaching an altitude of 48,000 feet. All six people on board lost their lives.

William Payne Stewart was born in Springfield, Missouri on 30 January 1957, the third child and son of a travelling salesman who specialised in beds and their accoutrements. Payne's father was a gifted amateur golfer who enjoyed the game and won many competitions at a level that, in 1955, saw him qualify for the US Open. Payne grew up in a loving family environment, brought up in the United Methodist faith, which he held to and was guided by, all his life.

It is not surprising to find that by junior high, Payne took a serious interest in golf and displayed some signs of what was to come. In his father, he found a good and demanding teacher, not only for the techniques of playing but also for the traditions and etiquette that a sportsman in the golf arena ought to know – including how to be gracious in both success and defeat.

Payne told his father that he wanted to be a professional golfer and would try to get a university golfing scholarship to cover his tuition fees. His father backed him and they found just such an offer at Southern Methodist University in Dallas, Texas. His academic course was in business and real estate, something at which he did not shine particularly well, but knowledge he found valuable when he moved into golf course design and building in later life. By the time he had graduated in 1979, Payne was the captain of the university golf team and he had no doubt in his mind that he wanted to earn his living playing golf professionally.

In Florida, Monday, 25 October 1999 dawned bright and sunny. For Michael Kling (age 42), an ex-USAF flier with 4,280 flying-hours in his logbook and now a commercial pilot working for Sunjet Aviation, it was going to be a busy couple of days. He had 66 hours on the Learjet and thus was relatively inexperienced on that particular type. Known for his strong religious beliefs, in his work Kling was also known as someone who paid attention to detail and did things by the book. As was his habit therefore, he arrived at the Sunjet facility at Orlando-Sanford airport bright and early at 06.30 EDT, so that he would have plenty of time to prepare for his flight. He was to make a series of five flights over the next two days. The first was a short, visual flight rules (VFR) positioning trip from Orlando-Sanford (SFB) to Orlando International airport (MCO), Florida, a distance of about 30 miles. There, Kling was to pick up a party of executives from the Florida sports agency, Leader Enterprises Inc., and fly them to Dallas-Love Field airport (DAL) in Dallas, Texas. From there, later the same day, he was to fly the group from Dallas to William P. Hobby airport in Houston, Texas. The following day, he would fly back to Orlando International and then take the aeroplane back to its base at Sanford.

Kling's first officer/co-pilot for this trip was Stephanie Bellegarrigue, a lively, athletic young lady, aged 27, who lived for her flying and was popular with the airline's clientele – which often included celebrities. She had 1,751 flying hours in her log-book, including 99 hours on this aircraft type. Familiar with her captain's habits, she arrived at 06.45, assisting Kling with the aircraft checks and monitoring the fuelling. The fuel load today was 5,300lbs, giving a maximum flying duration of around four hours – plenty in hand for what was expected to be a two hours and thirty minutes flight. A portable soft drinks-cooler box and a snack-basket were taken on board for the passengers. Departure on the short first leg was at

PAYNE STEWART: 25 OCTOBER 1999

07.54 EDT and they were down at MCO sixteen minutes later, ready to load their passengers for the next leg.

The passengers were flying to Dallas for a meeting with Ted Blackard, a golf course developer, to discuss the construction of a new course in Frisco, Texas. On board were the top men from Leader Enterprises: Robert Fraley and Van Ardan and it was they who had set up the meeting and hired the jet. A late addition to the group was Bruce Borland, an architect specialising in golf course design who brought with him a sound reputation having been involved with the design of many famous courses associated with the Jack Nicklaus organisation. The final passenger was Payne Stewart, the world-renowned golfer, member of the recent victorious US Ryder Cup team; twice winner and current holder of the US Open Championship. Borland had intended to take a commercial fight to Dallas for the meeting, but at the last moment, he was invited to fly down with Payne, to get to know each other better. Fraley and Ardan were personal friends as well as Payne's business agents.

The new golf venue would be named after Payne Stewart and he would have a significant input in the design of the course, which was intended would become the spiritual home for golf at the Southern Methodist University – which was based in Dallas and was Payne's 'alma mater.' Involvement in this aspect of the game interested Payne and he was keen to develop it as a potential future income stream for himself. 1999 was already turning out to be a very good year for him. Currently ranked third in the prize-money list and in the top-ten of world golf rankings, he had also found a new momentum to his game.

Payne Stewart was an instantly-recognisable figure on the PGA Tour. His stylish club-swing and high-class determined game made him popular with both aficionados and spectators of golf all around the world. It was, however, his distinctive, flamboyant style of golfing attire that never failed to draw attention to him and gave him a *frisson* of eccentricity that spectators loved. Turning professional in 1979, Payne won his first major title at the PGA Championship in 1989 and went on to win the US Open in 1991 and again in 1999. In 1986, he led the US Open for a brief spell in the final round, but tailed off to sixth place at the close. That same year, he gained second-place finishes in four PGA tour competitions and had no fewer than sixteen top-ten finishes – the highest-achieving player on the PGA Tour that year. In 1989, he achieved a top-ten finish in yet another eleven competitions, winning prize-money of \$1.2-million ($\approx$ \$2.9 million in 2024). It was at Kemper Lakes golf course, Illinois, however, that Payne showed his true mettle when he hauled himself from trailing six shots behind the leader to win by a whisker of one stroke, achieving his first title in a golf Major – the PGA Championship. In 1990, in a play-off, he picked up the MCI Heritage at Harbour Town in South Carolina for the second time – becoming the first player to win it back-to-back. By the close of the 1990 season, he was ranked as the fifth-best golfer in the world.

FATAL FLIGHTS OF THE RICH AND FAMOUS

Competing regularly in at least a couple of dozen tournaments a year, Payne Stewart's star was certainly rising and shone brightly when he came from behind to win his second Major title: the 1991 US Open at Hazeltine National course, Minnesota. At the 1993 Open Championship at Royal St George's in England, Payne shot a record-equalling 63 for the final round but poor earlier rounds let him down and he only managed twelfth spot.

Drawing on Christian teachings and his own faith, seemed to help him recover his competitive 'edge' once more and this became most evident during 1998, which saw a change for the better in his game and his earnings topped over $1 million that year (≈ $1.9 million in 2024). His big moment came in the US Open of 1999 at the Pinehurst course in North Carolina, when he sank a 15-foot putt to beat Phil Mickelson by one stroke and take the coveted title – his last Major. His much-photographed celebration skip-of-delight became remembered the world over and that joyful pose was the model for a statue of him erected after his death.

His record in the Ryder Cup, that fierce contest between the United States and Europe, should not be overlooked, either. Payne was selected for the US Ryder Cup team in 1987; 1989; 1991; 1993 and 1999.

The 1999 Ryder Cup was not the best playing week that Payne had ever had; his 0 wins-2 losses-1 halved, said it all, but he was a great team player and hoped, one day, to become a Ryder Cup captain himself. The US won the Cup and he felt so strongly about team-bonding that, after the tournament, he even dragged a youthful, exhausted Tiger Woods from taking a nap. Recalling that day, Tiger said: '[Payne] came into my room, woke me up and said: "get the hell out of bed and get your a** down to the team-room and experience what it is like to be part of a Ryder Cup team."'

All that was just a month ago. Invigorated by his successes this year, Payne was due to play in the Tour Championship, scheduled for 27-31 October at Champions golf course in Houston, to where he would fly with the others after the meeting in Dallas. At that point, he intended to remain in Houston; Fraley would go on to an appointment in Los Angeles and Ardan and Borland intended to return to Orlando.

On 25 October, the Learjet – registration N47BA – left for Dallas at 09.20 EDT and its planned track from Orlando International would take the aeroplane northward over Cross City, where it was to make a left turn on a new course towards Texas at a planned cruise altitude of 39,000 feet.

At 09.21 EDT Stephanie, whose voice confirmed she was handling the radio, contacted Jacksonville air route traffic control centre (Jackson ARTCC) to report climbing through 9,500 feet for 14,000 feet. The controller responded with an instruction to climb and maintain 26,000 feet and received an acknowledgement '2-6-zero, Bravo Alpha.' At 09.23, the controller cleared N47BA to Cross City and then direct to Dallas and again he received an acknowledgement. The next transmission from the controller instructed N47BA to change frequency to another Jacksonville controller and Stephanie acknowledged this change.

At 09.27:10, Stephanie reported climbing through 23,000 feet and at 09.27:13 the controller responded with an instruction to climb and maintain 39,000 feet, which was also acknowledged by Stephanie, at 09.27:18, with the response: '3-9-zero, Bravo Alpha.'

Things turned ominous at 09.33:38, when the controller instructed the Learjet to change frequency to a controller handling upper-level traffic. This time there was no response from N47BA.

Over the next 4½ minutes, the Jacksonville controller called five more times, but received no response. Fearing the worst, the controller initiated an emergency that was reported to the Federal Aviation Administration (FAA), which took immediate action. It was 10.00 EDT when the FAA contacted the military North American Aerospace Defense Command (NORAD) at its control centre in Colorado Springs and asked for assistance. This was not an unusual request in the circumstances and basically, the FAA explained it had an un-responsive aeroplane and asked NORAD if it could send a fighter up to take a look at it. There might be nothing sinister in it; maybe just a radio failure, but the Learjet had also not made its scheduled turn west towards Dallas and – more worryingly – west of Ocala City, about half-way between Orlando and Cross City, at an altitude of 32,000 feet, it had made an almost imperceptible six-degree turn towards the north. Now it was holding steady on a course of 320 degrees. The first response to the emergency was to scramble two ANG F-16s from Tyndall AFB in Florida. They were airborne at 10.10 EDT, just two minutes after the call but, when NORAD realised that the F-16 from Eglin was already airborne and could intercept the Learjet much sooner, the Tyndall fighters were recalled to base.

When Captain Chris Hamilton's F-16 caught up with the Learjet over Memphis, Tennessee, at 10.54 EDT, he found it flying along serenely at 46,400 feet, a thousand or so feet above the manufacturer's recommended maximum. It was now well off-course and being closely monitored by military units at NORAD; by the 1st Air Force at Tyndall AFB, Florida and by civilian ARTCC. 1st Air Force was the organisation responsible for the routine air defence of the whole United States, providing this through the US 'citizen air force' – units of the Air National Guard (ANG). The North American Aerospace Defense Command (NORAD) at its centre bunker in Colorado was responsible for handling potential air threats to the USA and Canada; initiating responses and liaising with the respective governments – and this biz-jet was heading Canada's way! A USAF Boeing E-3 airborne warning and control aeroplane (AWACS) and an ANG air-to-air tanker were also diverted to the incident. By now, the media, too, had sniffed out the story and TV news programmes ran 'flash' updates on the progress of the Learjet as it flew across America. The Pentagon and White House were kept informed of developments and there was much media speculation about whether the biz-jet would be shot out of the sky. Government sources claimed later that the escort fighters were not armed and since the projected flight path was over sparsely-populated areas and un-wavering, the question of taking terminal, armed action – which would have required President Clinton's approval – did not arise. A different view would almost certainly have been taken had the jet veered towards or been projected to fall upon a major conurbation – and the Canadian military also indicated they might take a stronger approach, too!

Meanwhile, Hamilton manoeuvred his F-16 all around the Learjet, as close as it was safe to go. The engines were running, the rotating beacon was on and there was no external damage visible, but the cockpit windows seemed misted or frosted over on the inside and his presence in such close proximity attracted absolutely no response. He was pretty certain these ominous signs indicated that the Learjet oxygen system had malfunctioned; the crew and passengers were dead and the aeroplane was flying on autopilot. At 11.12 EDT, having completed his inspection, Hamilton reported his findings and stayed with the 'ghost ship.'

Nothing happened for another hour until, at 11.59, Hamilton now low on fuel, proceeded to Scott AFB, Illinois. His lonely vigil was replaced at 12.13 EDT by two F-16s from 138th Wing of the Oklahoma Air National Guard (ANG), call-sign *Tulsa 13*, that were ordered to break off from an eight-ship training exercise and then directed to the Learjet by Minneapolis ARTCC. On arrival, these F-16 pilots: Lieutenant Colonel Mike Hepner and Lieutenant Colonel Mike Husted, also found the same situation. By now NORAD had received fuel-load data from the Sunjet management and it calculated the wayward jet might fall to earth in the vicinity of Pierre, South Dakota.

At 12.39, low on fuel, the *Tulsa 13* pair left to rendezvous with a KC-135 tanker from the 190th Air Refuelling Wing, Kansas ANG, based at Topeka, itself ordered to change course during a routine flight to Andrews AFB in Maryland and to take up station along the Learjet's track. All these aeroplanes were diverted to support what was now becoming a complex air mission – with what looked like only one outcome. Commercial air traffic, too, was being diverted as necessary, so that nothing came close to the track of the Learjet.

Hepner and Husted returned to formate on the Learjet, where they were joined at 12.50 by two F-16s from the 119th Wing of the North Dakota ANG, scrambled at 12.22 from Fargo AFB. Call-sign *Nodak 32*, these were flown by Major Kent Olson and Captain Rick Omang, who intercepted the Learjet over Sioux City, Iowa. The situation was reported to be unchanged and at 13.01, the *Tulsa 13* pair broke away to find the tanker again; it was a thirsty business.

What happened next was gleaned from the cockpit voice recorder (CVR) of the Learjet, which was one of the few intact objects recovered later. The CVR recording capacity was only thirty minutes, thus it would record sounds that were audible in the cockpit during the last half-hour of this flight. At this time, it was not compulsory to carry a separate flight data recorder (FDR) device in the USA.

At 13.10:41 EDT the CVR recorded the sound of an engine winding down. This was followed by sounds associated with a 'stick-shaker' (intense vibration of the pilot's control columns, designed to indicate the approach of an aerodynamic stall situation) together with an audible warning of auto-pilot disconnection and the continuous sound of the cabin altitude audible warning. At 13.11: 01, ARTCC radar detected the Learjet beginning a descending turn to the right.

One *Tulsa 13* F-16 left the tanker and followed N47BA down. *Nodak 32's* lead pilot reported the Learjet going down, performing multiple aileron rolls and apparently 'out of control, in a severe descent' – consistent with one engine stopped and no control surface movements in rectification. He, too, followed it down.

The CVR recordings stopped at 13.12:26 EDT.[1]

A catastrophic event affecting the cabin pressure and oxygen systems is believed to have occurred during the six-minutes between 09.33 and 09.38, at an altitude between 23,200 and 36,400 feet, north of Gainesville, Florida. The US National Transportation Safety Board (NTSB) investigated this crash but due to the almost total destruction of the aeroplane, there was little material on which the investigators could test their theories. The NTSB concluded that the probable causes of this accident to be incapacitation of the flight crew as a result of their failure to receive supplemental oxygen, following the loss of cabin pressurisation for undetermined reasons.

In the event, it turned out that the NORAD estimate of the impact point was not too far adrift, since the Learjet came down 2 miles west of Mina, which is roughly 100 miles east of Pierre, South Dakota. When the right engine stopped, a right-turn descent was detected by radar and the F-16s reported that the jet did rolls before falling into a spin all of which, from an altitude of 45,000 feet, could have propelled the Learjet closer to Mina.

It is not intended here to attempt to evaluate or comment upon the theories discussed in the NTSB report, since they are complex and beyond the scope of this account of the incident. Nor, for similar reasons, is it intended to comment upon the outcome of legal proceedings that arose in subsequent years from this sad event. Six people lost their lives as the result of a tragic accident. All of them – the famous and the unknown – were a grievous loss to their families, colleagues, friends and the general public.

Payne Stewart was and is commemorated in many ways. On the eve of the 2000 US Open held at Pebble Beach, competitors gathered to remember Payne as the first in their professional ranks who could not defend his US Open title. Eighteen of them lined up to hit eighteen golf balls into the sea in the golfing equivalent of an eighteen-gun salute to one of their own. There is a statue near the eighteenth hole of Pinehurst No. 2 course in North Carolina. A memorial stone has been erected at the crash site near Mina, South Dakota. A golf club, opened in 2009 in Branson, Missouri, is named after Payne Stewart. A section of the interstate highway 44 which passes through Springfield, Missouri has been designated the 'Payne Stewart Memorial Highway'. A statue was erected to Payne at Waterville Golf Links in Ireland.

Chapter 21

Hansie Cronje

In his short life Hansie Cronje rose to fame, first and foremost, as an international cricketer of quality but also, less creditably, as a self-confessed match-fixer. Sadly, in the world outside South Africa, he is remembered for the latter more than the former. In his native land, though, Hansie is revered as the highly-successful national cricket captain who lost his way and paid a heavy price – some say: unfairly – for doing so. Repercussions of his actions sent shock-waves through the cricketing world, where reverberations are still being felt to this day. His time in the limelight spanned just ten short years between 1992 and his death in 2002. During that time, he played his debut match for his country in the Cricket World Cup of 1992, then rose to captain South Africa in 53 Test Matches and 138 One Day Internationals (ODI).

Wessel Johannes Cronje, known universally as 'Hansie,' was born on 25 September 1969 to his parents Nicolaas Everhardus and Susanna Maria, in Bloemfontein, Orange Free State (OFS; known as Free State since 1995) province; the heart of the traditional, Afrikaans-speaking, Afrikaner region of South Africa. He was the middle of three children, with Frans as his elder brother and Hester his younger sister.

On the strength of his batting performance and leadership qualities in the South African Schools team, in 1988, at the age of 18, Hansie was selected to play for the Free State cricket team. Then, aged 21, Hansie Cronje made his debut on the international stage shortly after his country re-emerged from the cricketing shadows. It was the first test match following the ending of the ban when South Africa, led by Kepler Wessels, toured the West Indies in early 1992.

In their first game at the 1992 World Cup, South Africa shocked the hosts by pulling off an unexpected and impressive nine-wicket win over Australia in Sydney on 26 February 1992. Hansie was not required to bat but put in a tidy bowling performance of 5 overs-1 maiden-17 runs-0 wickets (5-1-17-0). He went on to play in eight of South Africa's nine matches. They enjoyed good wins over West Indies, Pakistan and Zimbabwe and reached the semi-finals, playing England in Sydney on 22 March.

In the years that followed, South Africa enjoyed regular success against India and Hansie was in good form for the Indian visit in the 1992/93 season. Here was a chance to witness the talents of Tendulkar, Azharuddin, Amre, Kapil Dev and

Manjrekar. Hansie was not at all over-awed, setting out his stall in the opening One Day International (ODI) at the Newlands ground in Cape Town, with a career-best bowling performance of 10-0-32-5 and one catch, and taking the player of the match award.

At the age of 24, Hansie was the youngest player in the South African national side and his great opportunity came when, as vice-captain on the tour of Australia between December 1993 and January 1994, he took over the captaincy when Kepler Wessels broke a finger during the second test in Sydney.

From February to April 1994, the Australians toured South Africa. Wessels was back as captain and Cronje enjoyed a particularly fruitful season with the bat. In six matches played over two weeks, Hansie hammered the Australian bowling for 721 runs. In front of his home crowd at Springbok Park, Bloemfontein, Hansie scored a magnificent career-best 251 off 306 balls in the OFS second innings. It was reckoned he was one of the best players of spin in the world. In October 1994, in a tri-series against Australia and Pakistan, he ended the tournament with 354 runs at an average of 88.50. At Centurion ground in 1997/98, he made 50 off 31 balls – what was then test cricket's third-fastest half-century – doing it in impressive style, with ferocious slog-sweeps, played on one knee and completing his 50 by pasting Muralitharan, the world's best spinner, for 4-6-6-6 in successive balls!

It was Kepler Wessels' last match as captain and Hansie took over permanently for the 1994/95 home series against New Zealand. Seen at this stage of his career as an adventurous captain, not afraid to take a risk, working with coach Bob Woolmer, these two men masterminded the course of South African cricket during the years leading up to the 1999 World Cup.

Between 1996 and 1998 Hansie never really produced the same quality in his batting. Although scoring 78 against New Zealand and 45 not out against Pakistan in the 1996 World Cup, he struggled to make a fifty during the next two seasons. Furthermore, since grabbing a five-wicket haul in an ODI against Zimbabwe back in 1995 and taking the key scalps of Inzamam-ul-Haq and Moin Khan in the third test of a tour to Pakistan in 1997/98, his bowling was pretty average, too. He regained his form during the tour of England in 1998, with five 50s on the trot; 81 in his win at Lord's; 69 not-out at Old Trafford; 126 and 67 at Trent Bridge and 57 at Headingley, to become South Africa's top-scorer with 401 runs at an average of 66.83, but sadly losing the series 2-1.

On home soil, Hansie and his South African team pulled off a white-wash in the December 1998/January 1999 test series against the West Indies, winning all five test matches by good margins, together with winning the seven-match ODI competition by 6-1. They followed this up with wins on a short tour of New Zealand in March 1999. It was in the cricket ODI World Cup of 1999, held in England, that Hansie made only 98 runs in total and South Africa suffered great disappointment in their famous tied semi-final against Australia at Edgbaston – the first tie in World Cup history and one in which Hansie made a duck when he was caught off his boot. There was another first, too, in this match when Hansie and Allan Donald took to the field in their match against India, each with a radio ear-piece taped into an ear. This gadget enabled them to receive real-time instructions from Bob Woolmer and was actually not against regulations, although the rules were somewhat vague about this technological aspect. TV commentators soon spotted the devices and the matter was brought to the attention of the umpires by Sourav Ganguly, who opened with Tendulkar. The umpires consulted the match referee, who in turn consulted the ICC representative, who decided that, although it was not strictly outside the rules, it was considered as 'unfair play' and the offending items must be removed. The ICC subsequently officially banned the use of such devices, but Hansie and Woolmer remained quite unrepentant about their actions. South Africa won the game but it left a bad taste.

Hansie Cronje was always regarded as a thoroughly charming, charismatic, inspirational character, by his fellow team-players, his contemporaries across the cricketing world, by commentators of the game and by his adoring home crowds.

From testimony given by Hansie during an investigation in 2000, the first indications of him overstepping the line in respect of dealings with betting and match-fixing, can be traced back to an ODI against Pakistan, played in South Africa between December 1994 and January 1995. South Africa hosted a four-way series with New Zealand, Pakistan and Sri Lanka, in what was known as the Mandela Trophy competition. An Indian or Pakistani book-maker is alleged to have offered Hansie $10,000 for his team to throw the game in Cape Town.

It can be seen, though, as the start of him taking the rocky road downhill since, having accepted tainted money, he developed an appetite for easy money and at the same time, he had been entrapped, exposing himself potentially to blackmail – or worse. It did not stop there, either. The same match-fixer now made a more tempting but, for Hansie, an even more risky offer. Now, he was being offered US$200,000 (≈ £340,000 in December 1996 or ≈ $390,000 or ≈ £657,000 in 2024) to get his team to throw the last match and only ODI of the tour in Mumbai on 14 December 1996. It seems truly amazing that Hansie was actually prepared to involve his team in this scheme. The lure of big money appears to have broken through the veneer of what the game of cricket was historically claimed to be all about – fairness, sportsmanship and the best team winning. But, was this, in fact, simply evidence that, despite public rhetoric and gestures, the game was no longer about such idealistic sentiments?

Hansie admitted that, before the match started, he talked to individual players and the whole team about this earth-shattering proposal. Collectively the team rejected the offer, apparently with some players speaking very strongly against it but, worryingly, others expressed an interest. Hansie dug himself further into the hole by bargaining for the offer to be increased and he and the book-maker settled on $250,000. On the morning of the match, he discussed the matter again with the team, but they turned him down again. Hansie scored 10 runs and the Indians won the match. He had burned his bridges by talking to the whole team; it was no longer a private matter. Would they keep it to themselves, though?

His unpredictable streak was further demonstrated during the fifth test, of the 1999-2000 series against England, played from 14 to 18 January 2000 at the Super Sport Park in Centurion, near Johannesburg. By this stage of his career, Hansie had lost much of the flamboyance and risk-taking swagger that had helped put South Africa back on the world cricket stage; had made him 'a blue-eyed boy' and a force to be reckoned with, in the early years of his captaincy. In the last few seasons, he had gained a reputation for being more defensive and risk-averse in his approach. It came as a surprise, therefore, when he made the England skipper, Nasser Hussain, an unprecedented offer. The outcome of the events that unfolded, would – eventually – go down in infamy.

On 14 January 2000, the first day, England won the toss and put South Africa in to bat. They scored 155 for 6 before rain stopped play. Then the heavens really opened and it rained heavily, making play impossible on days two, three and four. The match looked like a complete dead duck. Until, that is, Hansie came up with his masterpiece.

On the morning of the final day, the weather cleared. Hansie bumped into Alec Stewart and asked him if he thought Nasser would go for a deal to make a decent game of it. Hansie suggested a target of 270 runs in 73 overs.

What no-one realised was that bringing about this magnificent gesture in the interests of the game, was all a charade. Hansie Cronje was up to the hilt with the match-fixing fraternity and this was just another money-making scheme. His tormentors had persuaded him, the night before, to make his offer of saving the game, to Nasser Hussain in return for SA Rand (R)53,000 (≈ £5,000) plus – notoriously- a designer leather jacket for his wife.

Between 19 February and 19 March 2000, the South Africans followed their English series with a tour of India, consisting of a two-test match series and five ODI games and money and match-fixing reared their ugly heads again.

This money-mania did not stop. Before the second test in Bangalore on 2 March, Hansie was at it again. This time he suggested to Mark Boucher, Jacques Kallis and Lance Klusener that they could make some money if the match was thrown. They, too, put it down to Hansie's practical joking and refused to have anything to do with it. South Africa won that test by a big margin of an innings and 71 runs. It was in the final match of the ODI series at Nagpur that Hansie made an offer that lured two of his youngest players into his nefarious scheme. He asked Herschelle Gibbs and Henry Williams – and possibly other players, too – to play to order, in return for US$15,000 each, but it all went wrong. Gibbs was supposed to get out for under 20 runs and Williams was to have more than 50 runs knocked off his bowling. Gibbs claimed later that he forgot all about the plan and scored 74 before being run out and Williams went off injured before he had bowled two overs. As if that was not a big enough mess, it later emerged that Hansie had re-negotiated the deal with the bookie, for US$25,000 for each player, allegedly so that he could cream off US $20,000 for himself.

On 31 March 2000, in Sharjah, United Arab Emirates (UAE), Hansie led the South Africans out against Pakistan in the final day/night game of the Coca-Cola Cup series, a tri-country ODI tournament between South Africa, India and Pakistan. Hansie, showing all his aggressive brilliance, sparkled with a top-scoring knock of 79 runs off 73 balls, hitting six fours and three sixes, before being caught by Younis Khan off Arshad Khan's bowling but, despite decent scores by Neil McKenzie (58) and Mark Boucher (57), it was not enough to win the match.

It was the last game of first-class cricket that Hansie Cronje played and his world went into free-fall shortly afterwards.

On 7 April 2000, following the bugging of phone calls, on the basis of tape-recorded conversations police in Delhi, India charged Hansie with 'cheating, fraud and criminal conspiracy relating to match-fixing and betting' during the ODI series in India that March. Four other South African players were also alleged to be implicated. Some, but not all, transcripts of the conversations alleged to be about fixing the results of matches for money, were published, but the tapes themselves were not released. After public denials of all the accusations made against him and receiving support on all sides, Hansie rang UCBSA managing director Ali Bacher at 3.00 am on the 11 April to admit his dishonesty. It was a bombshell. He was immediately suspended from the national side, pending an investigation and Shaun Pollock took over as captain.

Convened on 7 June, the King Commission, an inquiry conducted by Justice Erwin King, was set up to examine the scandal and the vast scale of the matter was revealed.[1] On 10 June the Commission offered Hansie immunity from prosecution if he made a clean breast of his role in match-fixing. He agreed to this. He admitted to 'having a great passion for the game and for my team-mates that was matched by an unfortunate love of money' and taking large sums of money for giving information to book-makers, but claimed he never threw or fixed a match. The huge sums of money alleged to have changed hands – high six-figure amounts of US Dollars are bandied about – too, are unlikely ever to be truly clear; such payments being reported variously in SA Rand, US Dollars, UK Pounds Sterling or all of these and then often reported differently for the same event. Some of the book-makers were caught, some escaped and some were tracked down in later years. In 2003 there was a further revelation that seventy bank accounts with connections to Hansie, containing undisclosed sums of money and undeclared to South African tax authorities, had been tracked down to the Cayman Islands.

On 11 October 2000, on the basis of the damning evidence presented to the King Commission, the UCBSA banned Hansie Cronje from cricket for life.

HANSIE CRONJE: 1 JUNE 2002

The most frequent question asked is 'why did he do it?' It is unlikely the true answer will ever be known. In December 2000, Hansie's legal team mounted a campaign to have his ban lifted but the King Commission was closed down without a decision being made. In September 2001 Hansie appealed to the Pretoria High Court to have his ban removed, but that appeal was rejected, although he was permitted to coach privately and could practise as a journalist by attending matches as a spectator.

In order to carve out a new living for himself, he completed a Master's degree in Business Administration (MBA) and in February 2002, was taken on in a financial post by Bell Equipment, an international earth-moving machinery constructor located in Boksburg, east of Johannesburg. He usually dealt with insurance matters, but his experience in the public arena and his name came in very useful to Bell, when he organised and ran publicity engagements and social responsibility events for the company – and he was good at that work. He could also still command substantial fees for interviews by the world's media.

Home for Hansie was a four-bedroom thatched house that he and Bertha had had built in 1999 on Fancourt Estate, an exclusive golf complex in the remote small township of George, in Western Cape province. George was 650 miles south of Johannesburg but part of his employment package included Bell Equipment paying for a scheduled return-ticket to George and back – every weekend if he wanted.

In the evening of Friday, 31 May 2002, Hansie drove back from a business meeting in Swaziland (re-named Eswatini in 2018) and was slowed down by heavy traffic and heavy rain in the suburbs of Johannesburg. As was his habit, he planned to take a scheduled South African Airways flight to George, Western Cape to spend the weekend with his wife, but first he needed to call in to the office at Bell Equipment. This delayed him further and there was no way that he could make his scheduled flight from Johannesburg airport. It was cold and hailing when he rang a contact at AirQuarius, a small air-charter company to ask for a passage on their mail flight out of Johannesburg via Bloemfontein, bound for George airport. AirQuarius had a contract to provide a regular mail and freight run from Johannesburg to George. Hansie had a standing arrangement with his good friend, the owner of AirQuarius, that he could hitch a lift with them whenever he wanted. The carrier made no charge for the ride but in return, Hansie had a deal that he would provide the aircrew with stop-over accommodation in his spare room at Fancourt.

It was agreed that Hansie could have a seat on the flight, due to depart Johannesburg airport at 02.00 on 1 June. Hansie drove from his Boksburg office to the airport. Waiting for him was a small twin-turbo-prop airliner; Hawker Siddeley (or BAe) HS-748-372 Series 2B, registration ZS-OJU and its two-man crew, Captain Willem Lodewyk Meyer, a 69 years-old highly experienced pilot with 20,963 flying hours to his name (1,819 on the HS748) and his First Officer Ian Hughes Noakes, aged 50, with 1,100 hours. In 1986, this 748 was registered as N748AV operating with Air Virginia. It was sold in 1987 to SATA in the Azores as CS-TAP before being sold to a company in the UK as G-BICK. It was in 1999 that it finally moved to South Africa as ZS-OJU. The aeroplane would land at Bloemfontein airport after the first leg of 1 hour 37 minutes, to off-load some freight, then continue to George airport, where the remainder of its cargo would be off-loaded and where Hansie's wife Bertha would be waiting with a car to drive him and the two crew to Fancourt.

The HS748 took off from Bloemfontein around 04.00 for the two-hour leg to George. Weather conditions over the coastal destination were poor, with low cloud, rain and strong winds requiring the pilots to let down on instruments and effect a landing on runway 29 (heading 294 degrees), using the airport Instrument Landing System (ILS).

The approach to the runway was too high and too fast, so the pilots aborted the landing and followed a 'missed-approach' procedure prior to another attempt. This is standard practise and there is a set, documented procedure to follow that is specific to every airport. In the case of George airport, a missed-approach required the pilot to climb to 3,600 feet altitude on a heading of 294 degrees for a distance of 8 miles, then make a left turn towards the sea. At that point, with the approval of Air Traffic Control, the pilot could re-enter the landing pattern and try again. There was evidence that the ILS system on runway 29 was only working intermittently – it was plagued by faults – and may have failed during the first approach. There was a suggestion, too, that the standard missed-approach procedure was not followed. In working their way to re-position for a new landing approach, the absence of ILS signals, stronger than expected winds from the south-west and the aeroplane's directional gyro being faulty, combined to cause the pilots to lose situational awareness and the aeroplane to drift off-course. Unknown to the crew, the aeroplane was heading towards mountainous terrain. No external radar assistance was available at this airport; and audible height-warnings from the on-board ground proximity warning system (GPWS) – thirteen warnings were heard on the cockpit voice recorder (CVR) – were apparently not actioned.

At 05.15 on 1 June 2002, HS748, ZS-OJU crashed catastrophically into the Outenique mountains at Vandalenskloof ravine, 7.8 miles north-east of the airport and all three occupants died. The SACAA took until May 2004 to issue its report on the crash. The probable cause was given as 'loss of situational awareness' for the reasons mentioned above. Three recommendations were made: better maintenance of HS-748 instrumentation; replacement of the aging ILS apparatus on both runways at George; consider installing a ground radar system to assist landings at George. By then, the ILS had been replaced. It took even longer – August 2006 – to hold an inquest into the death of Hansie Cronje. In view of his dealings with unscrupulous, shadowy people, many conspiracy theories hung over his fate. However, held in the Cape High Court under Mr Justice Siraj Desai, it was the view of the court that the death of Hansie Cronje was brought about by an act or omission *prima facie* amounting to an offence on the part of the HS748 pilots.

Chapter 22

Vittorio Missoni

4 January 2013

When Vittorio Missoni, Chief Executive Officer and co-owner of one of the most successful luxury fashion houses in the world, disappeared in an aeroplane off the coast of Venezuela on 4 January 2013, initial thoughts turned to the possibility that he and his companions had been kidnapped for ransom. In the months following his sudden disappearance in this volatile South American corner of the Caribbean, it began to dawn on his worried family that Vittorio and his friends had most likely suffered a far worse fate. It will become evident, too, that this unusual story involves not just one incident but two that are inter-linked and that, eerily, both events occurred in the same area, on the same calendar date – 4 January – five years apart.

The Missoni company began life in Italy in 1953, established by Ottavio Missoni and his wife Rosita Jelmini and today the company – still regarded as a family business – is considered to be the master for Italian knitwear, while the brand enjoys equal status to Italian contemporaries such as Prada, Gucci, Versace or Ferragamo. Initially, Missoni manufactured and retailed knitwear for sports, leisure and social wear, in the Italian town of Gallarate before moving to Varese, near Milan. Vittorio Missoni, the eldest of three children, was born in 1954, the other siblings being his younger brother Luca, born in 1956, and his sister Angela, born in 1958. Being a close family, the founding parents handed over the running of the highly successful company to their children, around 1996 and in due course, Vittorio became the Marketing Director then CEO, Angela was the Creative Director and Luca the Technical Director.

Known for their bold designs, often incorporating geometric zig-zag and stripe patterns, it is as a result of Ottavio and Rosita's flair and innovative designs; excellent marketing; high-quality photography and regular exposure in trend-setting fashion magazines, that Missoni designs caught the public eye and became extremely popular the world over. Missoni couture became sought after and was – and still is – avidly showcased by the world's celebrities. In the early days of the 1960s and 1970s, having extended the collection to include a menswear range, clients included Hollywood icons such as Lauren Bacall and Marcel Mastroianni and the world-famous ballet star Rudolf Nureyev while, in more recent times names such as Catherine Middleton (aka HRH the Princess of Wales); Kate Moss; Beyonce; Gigi Hadid; Rhianna; Dua Lipa *et al*, have all been associated with the Missoni brand.

It was during the late-1990s and early 2000s that Vittorio guided the brand into perfumes then houseware, highly influenced by its use of the brand's vibrant colours and a sense of home and family. Aided by Creative Director Angela, he also entered the world of hotels with the opening of Hotel Missoni in 2009, a 136-room modernist building that lit up the Royal Mile in Edinburgh. It became the epitome of Missoni style and elegance and a showcase for the company's homeware range. A similar hotel was opened in Kuwait in 2010. He further extended the company's range into casualwear lines, such as M. Missoni, that were aimed at a younger, less-affluent clientele. Such was the success of the company that in 2023 its revenue was reported to be in the region of US$135million.

It was in late-December 2012, that Vittorio, aged 58, his wife Maurizia Castiglioni and four of their friends: Guido Foresti, and his wife Elda Scalvenzi, flew to Caracas in Venezuela and then on to Los Roques, for a holiday. Guido was the owner of a company that supplied and distributed construction materials, based in Pralboino, Brescia, Italy. He was well-known in motor-racing circles and owned a collection of classic Italian cars, some of which he raced in events such as the classic Mille Miglia. Also accompanying them was Eldas's brother, Guiseppe Scalvenzi and his wife Rosa Apostoli and they celebrated New Year 2013 in the remote island archipelago of Los Roques, a Venezuelan territory, designated as a national park, that lay about eighty miles off the coast of Venezuela, roughly north of Caracas. Los Roques was a bolt-hole for the affluent elite of Venezuelan society; regarded as a place to escape from a difficult political and sometimes dangerous, way of life on the mainland. Consisting of over 300 islands, islets and cays covering about 40 square kilometres, formed from the highest quality coral reefs around a large lagoon; it was remote; a sparsely populated, unspoiled, out-of-the-way, tranquil haven – that a person had to go out of his or her way to even get there. This became the attraction for high-wealth travellers from across the world and the islands transitioned from that of a subsistence fishing community to an economy based on tourism.

Enrique Perella is a Venezuelan aeroplane photography enthusiast with a group known as SVZM Spotters and he wrote of his own 'discovery' of Los Roques in 2006.

'My first time to come to Los Roques I was stunned by its unique beauty. The airport tarmac is pretty rough, but all kinds of private props [propeller aeroplanes] come every day. [It] became my favourite place on earth.'

On 4 January 2013, refreshed from their holiday in this idyllic Caribbean coral-reef playground of the wealthy, four of the party, Vittorio, Maurizia, Guido and Elda, were flying from the airport on the main island of Gran Roque, back to Maiquetia-Simón Bolivar International airport in Caracas on the mainland of Venezuela. Guiseppe and Rosa, however, had decided to remain a little longer on the island to enjoy more of the wonderful climate. Therefore, they did not board the aeroplane and said they would catch another flight back to Caracas later – which indeed they did. Since, at this time, no large airliners were able to operate from the island's small runway, this was to be just the first leg of the four passengers' journey back to Italy. It was only a short hop of eighty miles and the party had hired an aircraft and crew from a local air charter company called Transaereos 5074.

It is not clear how and with whom the flight was arranged and details of the flight company; its employees and the transaction are almost non-existent which, unfortunately, tends to fit in with a veil of vagueness, anomaly and speculation that envelops the air component of this story. It is alleged that Transaereos was not licenced to conduct ticketed, commercial flights and that such flights were done on an *ad hoc*, private basis, with no oversight by any regulatory body. There have been some, unsubstantiated, suggestions in the media that aircraft maintenance was not all it should be either. The aeroplane allocated for the trip was a 45-years-old British designed and built Britten-Norman BN-2-27 Islander, with the Venezuelan civil registration: YV2615. Journalists, with some difficulty, later established that the owner was a gentleman from Caracas they named as Asdrubal Remigio Bermudez Gonzalez. He was alleged to have operated an airline in the Venezuelan state of Miranda for a about a year before YV2615 turned up in Los Roques, just a few months prior to this incident. The pilot for this particular flight was Hernan José Marchan, said to be aged 72 and his co-pilot: Juan Carlos Ferrer Milano, aged 45. There are almost as many different spellings of the pilot and co-pilot's names as there are press reports, but the foregoing are the most-quoted versions. It is reported that Marchan's pilot licence had expired in November 2012[1] and that Gonzalez told a reporter, in what appears to be his only interview[2], that Marchan was a highly experienced pilot with many hours on the type and that the aircraft was serviced correctly.

In respect of what will become the principal aeroplane involved in this incident – for it is by no means the only aircraft relevant to this story – its life begins in the small Britten-Norman aircraft factory on the Isle of Wight, England. On 22 January 1968, airframe number 0020 of a model known as the BN-2A-27 Islander was rolled out bearing the new British civil registration: G-AWCB. The Islander was a very successful aeroplane for this small British company and 1,300 examples have been sold to date – with nearly half that number still believed to be in service around the world, in civil and military service – the latter, for example, known as Defender in current RAF service.

In April 1968, Britten-Norman sold G-AWCB to an American aircraft dealer: Alamo Aircraft Inc., in Texas, where it was given the new US civil registration of

N585JA. Alamo sold it to another dealer: Altus Airlines in January 1969 and later that year, it was bought by Colony Airline, which operated in the Bahamas. This was the third of some sixteen changes of ownership over the next forty years, to operators mostly located in the southern USA, the Caribbean area and the north of South America – although one exception was when it had a spell with the Oregon (USA) State Police between 1986 and 1988. During its time in the Bahamas, this twin-Lycoming piston-engine, high-wing, ten-seater (usually one or two crew plus up to eight passengers) aeroplane was employed on air charters and it was used to provide a daily, third-level air service from Nassau, on the island of New Providence, to Georgetown on the island of Great Exhuma, an over-water distance of 143 miles. Around 1971, when Colony Airline ceased operations, the Islander was sold to Pelican Airways, based in Florida and the registration mark was changed to a slightly confusing N555JA. It was photographed with that mark at Miami-Kendall Tamiami Executive airport in Florida in February 2009 and is then recorded in November 2009 as being sold to an unidentified owner and coming off the US list when it was re-registered as YV-2615 – YV being an aircraft nationality code for Venezuela. At the time of the fatal flight, Islander YV2615 was believed to be the fourth oldest example of a BN-2 still flying.

The doomed flight took off from Gran Roques airport runway 07 (oriented 076/256 degrees) at 11.32 local time. It made a climbing turn to the left, normal for a departure from Gran Roques *en route* to Maiquetia, and began its ascent to the planned cruising altitude of 6,500 feet. At 11.42, the pilot radioed that they were ten nautical miles (nm) from the Gran Roques VOR (a radio navigational aid, for pilots to confirm position and help maintain a course, known as: Very High Frequency [VHF] Omnidirectional Range), at 5,000 feet. The Gran Roque air traffic controller then instructed the pilot to change radio frequency and contact Maiquetia (Caracas) airport approach control. No such contact was established and no distress calls were made by the crew – or, at least, none were heard. Data collected later from the radar facility on the mainland showed that the aeroplane continued to climb. It reached an altitude of 5,400 at a distance of 13.2nm from the Gran Roque VOR and was flying at a speed of 120knots. It was at this point that the aeroplane appeared to lose speed and altitude rapidly while turning to the right, before disappearing from radar contact.

By first light the next day, a search was underway. Early in the proceedings, the Italian Ambassador to Venezuela flew over the area and Vittorio's brother Luca, himself a pilot, travelled from his business base in New York to represent the families and he, too, joined the search operation on board one of the search aircraft. Four aeroplanes, two helicopters, two coastguard ships, five patrol boats and some 400 personnel including thirty divers, were engaged in this rescue mission but, with no surface signs, the search was scaled back after about ten days. It was announced by government sources that, on or about 14 January, while walking among rocks on its eastern shore, a tourist on the island of Curacao in the Dutch Antilles discovered a backpack that was later confirmed to belong to another Italian holidaymaker – most likely Guiseppe Scalvenzi – present on Los Roques at the time of the accident. Apparently, this person had been told that his bulky bag – said to contain kite-surfing gear – was one bag too many and could not be accommodated on the aeroplane in which he was going to fly to Caracas, so he had asked the Missoni party to carry it on their aeroplane as a favour. Curacao is 135 miles west of Los Roques.

During February, two more pieces of luggage that were linked directly to the people on board the Islander, were washed up on the island of Bonaire, 95 miles west of Los Roques.[3] These finds caused the search pattern to be extended in that direction, but without further success. The actual search area is another unclear statistic, but Venezuelan government sources claimed a total area of 27,000 square miles of sea and coast had been searched.

Before this latest incident, personnel in the local SAR organisation, though professional airmen and keen to help, had limited resources available. When a twin-engine Beechcraft BE-80 Queen Air aircraft, YV-539C, operated by Viproca airline, was lost on 20 December 1997 with one pilot and nine passengers on board, the Venezuelan Air Search and Rescue (SAR) organisation, while very willing, was not well-equipped or organised to carry out speedy, methodical, electronically-aided air search missions. In a subsequent court case brought by the sole survivor, it was considered that a lack of suitable equipment, organisational confusion and poor response time, contributed to the high loss of life in that incident when, despite several passengers being seen alive in the water, only one was rescued. The aircraft suffered engine failure while descending into Caracas airport from Los Roques and it crashed into the sea about 13 miles offshore. As a result, the government reviewed its SAR capabilities and acquired three helicopters fitted with devices such as infra-red, heat-seeking equipment, which was a vast improvement when searching for survivors on the sea's surface.

In the case of the Transaven Let-410, YV-2081 incident that occurred in 2008, sadly those very helicopter resources were already tied up at the critical time on another mercy mission elsewhere on the mainland. Vital time was lost before they were available to conduct an over-water SAR mission. Despite the helicopters being rapidly deployed when the Transaereos Islander, YV-2615, disappeared in 2013, there was no surface evidence for its equipment to home in on.

One eerie coincidence between these last two incidents is that they occurred on the same calendar date of 4 January and there are other similarities, too.

On 4 January 2008 a Let L-410UVP-E3 Turbolet, YV-2081, operated by a small airline called Transaven – which was one of many that specialised in linking Caracas with the tourist islands of the southern Caribbean – was on a non-scheduled private flight from Caracas airport to the islands of Los Roques. The Let-410 is a twin-turbo-prop-engine, high-wing aeroplane capable of carrying up to nineteen passengers and two flight crew. It is manufactured by a Czech company: Let Kunovice (later: Aircraft Industries) and production to date has reached about 1,200 aircraft. YV-2081 was built in 1987.

It had two crew and twelve passengers on board, including eight Italian, one Swiss and five Venezuelan citizens, two of whom were the crew. All was going well until, about 12 NM south-south-west of the islands, the pilot radioed that both engines had stopped and the aeroplane was at 3,000 feet, descending rapidly. This was the last contact with the aircraft. It had not quite disappeared without trace since, when another Transaven aeroplane flew over the area a tiny slick of petrol or oil was spotted, but that soon disappeared, too. When a more comprehensive search was initiated, only one lifeless body was found on the surface by the crew of a fishing boat eight days later. This was the co-pilot: Osmel Alfredo Avila Otamendi, age 37.

Returning now to the missing Islander; after doing the best they could by air and sea, the Venezuelan authorities scaled back the official search in January after about ten days. It was clear that no further progress could be made without the aid of sophisticated underwater search apparatus and Venezuela did not possess such facilities.

At the end of January 2013, a Venezuelan government minister informed the media that in early February a US-owned deep-sea survey ship would arrive to enhance the search for the missing Islander aeroplane. Its search operation would be jointly financed by Italy and Venezuela. The company engaged was C&C Technologies Inc., working in conjunction with SEA Corporation. They despatched the research vessel R/V *Sea Scout*, a 134-foot aluminium oceanographic catamaran to undertake a search of the seabed. The vessel was equipped with an autonomous underwater vehicle (AUV) with an impressive array of sensors, imagery systems and cameras and experienced personnel.

Sea Scout sailed from Italy and arrived in Venezuela on 22 March but, due to technical problems with the AUV and the need for testing after repairs, the seabed search did not get under way until 19 June. However, by 27 June not one, but both missing aeroplanes had been located and identified.

Late in the afternoon of 19 June 2013, the first aircraft to be found was the Let L-410, YV-2081. Its exact location was kept under wraps but, next day, the Venezuelan Attorney-General announced that it was found south of the archipelago at a depth of 970 metres (3,200 feet). Photographs, later published on-line, clearly showed the prominent registration letters and numbers and that the aeroplane was lying upright on the seabed, quite intact, with no detachment of the main wings or tail assembly. This suggests the aeroplane probably crash-landed on the sea but sank before the occupants could escape. Having satisfactorily established the identity of the aircraft as that missing from five years earlier, in deference perhaps to the feelings of the relatives of the missing, who now had a form of closure, it appears there were no plans to further disturb what was in effect a burial site. The depth, too, is a very significant factor.

Using more of the mainland radar-derived co-ordinates, the *Sea Scout* crew now moved on to the area consistent with the last known position of the Islander and began a methodical search. In the evening of 27 June, the wreckage of BN-2 Islander, YV2615, was discovered north to north-west of the islands at a depth of 75 metres. Again, photographs confirmed the registration, but in this case the aeroplane was in a broken state. This suggests the Islander impacted the sea under power, with no chance of escape for the occupants. The depth of this site would, however, allow divers to inspect the possibility of human remains being present and for the recovery of wreckage to the surface.

Once away from the coral reef shelf of Los Roques, the sea depth increases rapidly. For example, it goes down to 2,000 metres at the bottom of the Bonaire Basin in the area to the south where the Let-410 was found and even deeper to the north, in the Los Roques Basin.

The Venezuelan SAR authorities closed their case and handed the next phase to the Navy to organise a wreckage and casualty recovery operation on the Islander.

On 1 July 2013, Angela Missoni issued a statement on Facebook, on behalf of the family. She said she was in touch with the Italian Foreign Ministry and other state bodies to officially report the disappearance of her brother Vittorio and was in Rome to co-ordinate information and the offers of aid that were being received from around the world. Her brother Luca, who is a pilot and the current Missoni company Chief Executive Officer, Alberto Piantoni were both in Venezuela taking an active part in the search and liaising with the Venezuelan authorities; the Unity of Crisis and the Italian Embassy. She expressed the family's thanks for all the affection received and efforts being put into the search and rescue operation and that the family would not give up. The Unity Advisory Group, mentioned here, is a private organisation that offers advice to businesses, high-net-worth families, governments and non-governmental bodies, on effective solutions in challenging and high-risk situations and environments.

On 17 October the government announced that divers had found human remains and that samples were taken for DNA analysis. There then followed a series of conflicting press reports about how many and who had been found, to such an extent that they are considered too unreliable to repeat here.

Wreckage from Islander YV2615 was finally brought to the surface on 25 November 2013 but from that point onwards a veil of silence descended and nothing further was reported in the press or by the family. The veil lifted slightly some ten years later when the Italian newspaper: *Chronicle Pralboino* reported that a mass was held in Pralboino to remember Guido Foresti and his wife Elda.[4]

In the case of the Let-410, the Venezuelan JIAAC (Junta Investigadora de Accidentes de Aviación Civil) authorities issued a formal report on its investigation but it did not contain any conclusion or recommendation. The report was also only published locally in Spanish which, since that is not the international aviation language of English, was not widely circulated and therefore of little benefit to the aviation world. In the case of the Islander incident, no formal report from Venezuelan authorities has been traced on internet sources as yet.

Chapter 23

Kobe Bryant

26 January 2020

In the minds of people of a certain age, the phenomenon that is modern basketball played on a media-hungry world stage, may not have progressed much beyond fond memories of the Haarlem Globetrotters and their mesmerising antics – particularly those of their wonderfully-named star: Meadowlark Lemon. In the USA, however, the sport of basketball is nothing short of a religion to its followers and as such it has created its own icons, whose statures have become known far beyond those shores. One name that stands, literally, head and shoulders above the rest is Kobe Bryant, who retired in 2016 after a distinguished twenty-year professional career spent playing entirely with the Los Angeles (LA) Lakers team.

In his sport, there is nothing that Kobe did not achieve. Every accolade that could be given, was given – and was richly deserved. His prodigious talent made him a rich and famous man and he bore that fame with dignity and consideration for others. It was while demonstrating his commitment to his family and to the next generation of young players, that he tragically lost his life in an air accident on 26 January 2020.

Born on 23 August 1978 in Philadelphia, Pennsylvania, Kobe Bean Bryant had basketball in his genes. Kobe was the youngest child and only son of the three children of Joe and Pamela; his elder sisters being Sharia and Shaya. His middle name was a nod to his father's own playing-days nickname. Kobe's father, Joe Washington Bryant (known in his playing days as *Jelly-Bean*) was 6 feet 9 inches tall; a professional who played eight seasons in the USA National Basketball Association (NBA) league, with teams such as the Philadelphia 76-ers; San Diego Clippers and Houston Rockets. Moving, in 1984, to take up a position with a team in Italy, Joe continued playing for another eight seasons for teams in Italy and France. Thus, Kobe had his early schooling in Italy and indeed became fluent in the language as well as taking up serious basketball under the tutelage of his father. The family returned to the USA in 1991, when Kobe was 13 years of age and his father turned his talents to basketball coaching between 1992 and 2015, in the USA and later, in Thailand and Japan. Kobe attended Lower Merion High School in

Ardmore, PA, where his own basketball career began its stellar trajectory. His break into major league basketball came in 1996 when he joined the Charlotte Hornets, who soon sold him to LA Lakers for the 1996/97 season, becoming the second youngest player in the history of the NBA. He crowned that season and set down a marker for his future, by winning the annual NBA 'Slam Dunk' competition. By the end of only his second season, the 6 feet 6 inches-tall, shooting guard Kobe was an NBC All-Star player, the first of his eighteen All-Star selections. The NBC All-Star game is an annual exhibition game; the main event of the NBA All-Star Weekend. Traditionally, the All-Star Game was a conference-based format, featuring a team composed of all the top ranked basketball players in the Eastern Conference against another team of all-stars from the Western Conference. Kobe played in this top match no less than eighteen times during his twenty-year career. Some of the basic statistics of his outstanding career with the LA Lakers is probably the most succinct way to convey the extent of his achievements.

Kobe played 1,346 games, in which he scored a total of 33,643 points at an average of 25 points per game. This is the fourth highest total score in NBA history behind LeBron James (41,261); Kareem Abdul-Jabbar (38,387); and Karl Malone (36,928). He was twice the top scoring NBA player (2006 and 2007); with LA Lakers he won the NBA Championship in five seasons; he was classified as the NBA's most valuable player in five seasons and was selected on no less than twenty-seven occasions to play in various prestigious All-NBA teams. He was also selected for the basketball teams representing the USA at the 2008 Olympic Games in Beijing and the London Olympics in 2012 – winning an Olympic gold medal at each.

Kobe was troubled by various injuries to a knee, Achilles tendon, right shoulder and calf during the years between 2013 and 2015 and these pressures were key factors during the 2015/16 season when he decided to call time while he was still at the top. His earning power as a player was prodigious, with contracts involving sums of $30 to 40million a year. It has also been reported that during his playing career he earned in the order of $680million in salary and commercial endorsements.

He married Vanessa Laine, a young lady raised in Anaheim and Garden Grove, in April 2001 after a six-month courtship and they raised a family of four daughters: Natalia Diamanté (born: 2003), Gianna Maria-Onore (2006); Bianka Bella (2016) and Capri Kobe (2019), who were adored by their parents and to whom Kobe – while juggling his super-high public profile – went out of his way to be an 'always-present' father to them all. Home for the Bryant family was a four-storey, 16,000 square-feet custom-built house with 'million-dollar views' of the Pacific Ocean. It was located in a secluded part of Pelican Crest, a large gated complex that was home to similarly high-wealth residents, on Newport Coast to the south of Los Angeles, where properties were being offered for sale in excess of $20 million in 2024.

While he was still playing, he wisely began planning for his retirement by setting up Kobe Inc., in 2014 in Newport Beach, from where he controlled all his business activities. It was a brand development operation that had Body Armor as its first investor for $6 million. In four years, that investment grew to a value of $200 million and became linked with Coca-Cola. He was involved with a venture-capital fund that invested in media and tech companies that grew to manage an alleged $2 billion capital sum. Kobe ventured into co-operation with a Chinese retail company, selling through social media platforms. Not least, he founded Kobe Studios (later Granity Studios), a multimedia production company based in Costa Mesa that promoted sports to paper, podcast and film outlets. It was this desire to be 'there for his family' that, on one occasion, prompted him to explain why he used helicopters so much to commute around Los Angeles – at an alleged $10,000 a flight. In a podcast, recorded in 2018 and recalled by the Huffington Post,[1] Kobe said he took helicopters around LA so that he could spend as much time with his family as possible. He frequently became stuck in the awful LA traffic, which sometimes caused him to miss picking up his daughters from school and enjoying after-school activities with them. He said the traffic was really bad and once, he was sitting so long in traffic that he missed a school play and felt really bad about it. He needed to come up with a way that he could train and focus on the basketball side, but not compromise his time with the young family. The commute from his home to the LA Lakers' facility in El Segundo and to home games at Staples Center in LA was so inconvenienced by the traffic-laden roads that he chartered helicopters as often as possible. That, he said, was when he looked into using helicopters to be able to get down to training and back in around fifteen minutes. That was how it all started. He said his routine then changed and he would drop off the children at school, fly down, practise hard, fly back and then get in the car-pool line and pick the children up again.

The two eldest children took up an interest in their father's sport at a very early age. Natalia became a good volleyball player before moving into the world of modelling. Gianna, very tall for her age, became an excellent junior basketball player at Harbor Day School in Newport Beach. She played for the school team and with encouragement from her father, was heading towards a bright future in the sport.

On Saturday, 25 January, Kobe took one of his daughters to the local shopping mall and then took another to a basketball game. The following day he planned a trip out with his daughter Gianna and some friends, who were to play in a basketball tournament. First, though, on that fateful Sunday morning, at around 07.00 Kobe stopped by his usual place of worship: the Catholic church of Our Lady Queen of the Angels, in Newport Beach. He did not stay for Mass, but went inside just for a few minutes to pray; on his way out, he met the parish priest, Father Steve Sallot and exchanged a few words with him, before heading off to the airport. Father Sallot recalled later that he knew Kobe had prayed because he: 'saw a drop of [holy] water still glistening on his forehead.'

The purpose of the trip on Sunday, 26 January was to accompany his daughter Gianna and two of her team-mates, Payton Chester and Alyssa Altobelli, to Kobe's

Mamba Sports Academy in Thousand Oaks and to give his support and coaching advice during the course of the game. The mere presence of Kobe Bryant in the stadium would be a terrific boost to the home side as well as boosting overall attendance. The three girls were to play in an Amateur Athletic Union (AAU) club basketball game against opponents: Fresno Lady Heat, in what was known as the Mamba Cup[2]. The intention was for the party to fly by helicopter from the nearby John Wayne-Orange County airport, Santa Ana to Camarillo airport in Ventura County, then be driven by car to the Mamba venue in Thousand Oaks, about ten miles from the airport.

The helicopter, a twin-engine Sikorsky S-76B, registration N72EX was built in 1991, the 374th to roll off the production line. This specific aircraft was previously owned and operated by the State of Illinois, with the registration of N761LL, and was one of four such helicopters used for VIP transport by the State Governor and his administration between 2007 and 2015. In 2015 the Governor, in an effort to reduce State expenditure, instructed that the four helicopters were to be disposed of. They were auctioned off by the State and with 3,951 hours on the airframe N761LL was sold to Island Express Holdings of Van Nuys for $515,000. Its registration was then changed to N72EX. During its use while Kobe Bryant was finishing his playing career, it had a black and silver paint scheme and bore Kobe's Black Mamba trademark logo. It was later re-painted in a white and blue scheme and in addition to being used for travel by Kobe Bryant, was also used for flights out to Catalina Island, general charters, air tours and holiday packages. It had a comfortable executive-style seating layout for up to twelve passengers and was provided, complete with its pilot, by Island Express Holdings Inc., in response to an advance booking by an air charter broker on behalf of Kobe. As explained above, Kobe was a frequent user of helicopters, particularly from this company and for example, N72EX had been hired out to him at least twenty-four times in the past two years – not always with the same pilot. The S-76 had provision for a crew of two but Island Express normally operated theirs with just one pilot, since their on-demand-type flights were generally made under Visual Flight Rules (VFR) conditions. On this occasion the single pilot was Ara Zobayan age 50; ten years with the company and currently its chief pilot. He had logged a total of 8,200 flying hours, 1,250 of which were on the S-76 helicopter type and Ara made himself available to fly Kobe whenever possible. He had been a licenced helicopter pilot since 2001 and obtained a commercial helicopter pilot and instrument-flying rating in 2007.

The full report of the US National Transportation Safety Board is drawn upon as a reliable source of information to follow the course of events as they unfolded that day[3].

While Kobe dropped by the church and made his way to the airport to meet his party, Ara Zobayan was up early, around 06.30, to check the weather and his aircraft, complete his paperwork and then to re-position the helicopter from the company's base at Long Beach/Daugherty Field in Long Beach CA, to John Wayne airport. This was a ten-minute hop and he arrived at John Wayne at 08.37.

It was an overcast cloudy day, with a cold front passing through Ventura County that pulled fog in from the ocean and up over the Santa Monica hills. Using a widely-used, integrated flight planning, information and hazard assessment app for pilots, called ForeFlight – accessible on his mobile phone and tablet – the pilot updated his weather data; completed his risk analysis and other paperwork for the flight and in conversation with the owner of the air charter broker, judged that the flight could proceed. Bearing in mind this trip was approved as an on-demand, single-pilot VFR flight (i.e. prohibited from entering clouds), weather conditions *en route* were variable, with low cloud and poor visibility in the mix. 'Blue' conditions, as designated by the planning app, were forecast (cloud ceiling 1,000 to 3,000 feet above ground level [agl] and/or visibility from 3 to 5 miles) for the Long Beach, Santa Monica and Malibu Pass sector. This deteriorated to 'Red' conditions, as indicated by the app, (500 to less than 1,000 feet agl and visibility from 1 to less than 3 miles) over the downtown Los Angeles, Burbank and Van Nuys sector. The 'Red' conditions were due to more cloud, but the pilot told his flight operations director that he intended to go up and round the weather on a course that would take him east and north of the clouds. He intended to fly north to the Dodger Stadium in LA; go round Burbank then pick up and follow State Highway 118 to his destination: Camarillo airport.

It later emerged that Los Angeles Police Department (LAPD) Air Support Division, due to low visibility conditions, had grounded all its own helicopters until late that afternoon. The Los Angeles County Sheriff department also grounded its helicopters for the same reason.

The people on board N72EX that day were:

> Kobe Bryant, retired basketball player (age 41).
> Gianna Bryant, his daughter (13).
> Payton Chester (13), team-mate and school-friend.
> Alyssa Altobelli (14), team-mate and school-friend.
> Sarah Chester, Payton's mother, (45).
> Keri Altobelli, Alyssa's mother, (46)
> John Altobelli, Alyssa's father, a noted Orange County College baseball coach, (56).
> Christina Mauser, assistant basketball coach at Mamba Academy, (38).
> Ara Zobayan, pilot, (50), resident of Huntington Beach, CA and chief pilot of Island Express Inc.

The helicopter took off from John Wayne airport at 09.07. The flight north-westwards proceeded at an altitude of between 400 to 600feet agl at a speed of 140 to 150 knots and for the first half-hour or so it was uneventful. At 09.20 the pilot was flying at 800 feet and 8½ miles south-east of Bob Hope airport in Burbank, when he contacted its tower air traffic controller. He requested Special Visual Flight Rules (SVFR) clearance to pass through Burbank airspace to allow him to follow the road US Route 101, which went west towards Camarillo. The pilot was asked to hold outside Burbank airspace, due to volume of air traffic being handled at that time, while his request was dealt with. He was also advised that cloud tops were at 2,400 feet above mean sea level (amsl). At this point, the helicopter circled south-east of Burbank for twelve minutes, in the vicinity of US Highway 101. With the weather declining, this was the point at which he had to decide to press on or to land at Burbank airport. He chose to continue.

At 09.39, his request was approved and his routing was now in the hands of an air traffic controller at Van Nuys airport, who directed him northwards around Burbank airport to keep clear of other air traffic. Heading towards the Granada Hills, Ara climbed from 800 feet to 1,400 feet then, at 09.39 he was instructed to turn southwest to intercept Highway 101 and to follow it to Camarillo. At 09.42 he was following a familiar route again, while heading towards the hills around Calabasas, which was covered in fog. He was flying just above the cloud tops but the clouds were rolling up, following the contours of the hills. On radar, controllers saw the helicopter climb from 1,250 feet to 2,125 feet at around 126mph and it is believed to have flown into the clouds rolling up over the steep hillsides. Probably, realizing he was in cloud over steeply rising ground, Ara was now advised by control, at 09.44, that he was too low to be seen on radar. He was now flying west at 450 feet agl (1,370 feet amsl) over Highway 101 and among rising terrain. He informed air traffic control that he was going to climb above the cloud layers and the helicopter began a climb at the rate of 1,500 feet per minute. The helicopter began a gentle left turn roughly following Highway 101 but half-a-minute later, still climbing, the turn became tighter and it veered away from Highway 101. His ground proximity warning system may have been telling him to 'Pull Up' and all this input would not be calming. His reaction might have been to pull up, but by then it is surmised that the pilot had entered a state of spatial disorientation which resulted in the helicopter plummeting to the ground. When the helicopter reached 1,600 feet agl (2,370 feet amsl) at 09.45, it began to descend rapidly towards the ground, but, in answer to a question from the controller as to his intentions, the pilot responded that he was climbing to 4,000 feet amsl. At that point, the aircraft was actually descending.

The helicopter was seen by a witness to emerge from the bottom of the cloud layer in a left-banked descending turn and crashed on a fog-enshrouded hillside in the district of Calabasas at 09.46. It burst into flames and there were no survivors.

The NTSB determined that the probable cause of the accident was the pilot's decision to continue the flight under Visual Flight Rules (VFR) into Instrument Meteorological Conditions (IMC), which resulted in the pilot experiencing spatial disorientation and loss of control. The NTSB investigators considered that a factor contributing to the accident was the pilot's likely self-induced pressure and the pilot's plan continuation bias, which it was felt adversely affected his decision-making. The report also highlighted the operating company's inadequate review and oversight of its safety management processes.

Given the circumstances, it was inevitable that Kobe's wife and family would initiate litigation against the estate of the pilot and the owners and operators of the helicopter. It is not appropriate to discuss this aspect in detail here. A lawsuit was filed in February 2020 and it was reported by the LA Times (and others) on 22 June 2021 that a settlement had been agreed in favour of the Bryant family; the terms of which were classified as confidential.

Sadly, this was not the only matter for litigation. Reports emerged in the media during late-February 2023 that Los Angeles County Board of Supervisors had agreed to pay Vanessa Bryant and her daughters a very substantial sum in settlement of a lawsuit brought against the County. This lawsuit alleged emotional distress as a result of Vanessa Bryant becoming aware that some employees of the LA County fire department and the Sheriff's department had shared among themselves graphic photos of the crash site including distressing images, allegedly taken while attending the aftermath of the crash. A similar case relating to the other victims had already been settled by the other families during 2021.

Kobe Bryant and his daughter Gianna were laid to rest in a private funeral on 7 February at the Pacific View Memorial Park in Corona del Mar, not far from the family home. A public memorial attended by thousands, was held for the other victims on 10 February, followed by a public memorial honouring not only Kobe and his daughter but all the casualties of that tragic accident. This was held at Staples Center, home of the LA Lakers, on 24 February and was attended by stars of the basketball world, past and present, together with showbusiness celebrities and thousands of ordinary people, there to remember the life of Kobe Bryant.

Appendix 1

Cierva

List of crew and Passengers aboard Cierva's DC-2, PH-AKL, 9 December 1936. Flying from Croydon to Amsterdam and onward to Berlin. (K=killed).

Crew			
Captain Ludwig Hautzmayer.	Pilot.	K.	Ace fighter pilot in First World War.
Jacob Verkerk.	Flight engineer.	K.	
Pieter Cornelis Valentjin van Bemmel.	Wireless operator.	Died of Injuries.	
Hilde Bongertman.	Stewardess.	Slightly Injured, survived.	KLM's chief stewardess. A member of the Dutch National Socialist Movement, she was prosecuted and interned for her wartime connections to the Nazi party.
Passengers			
Juan de la Cierva.	Spanish.	K.	Believed to be on a business trip to Berlin.
Charles Robert Dickson.	Swedish.	K.	A Swedish banker and sportsman. Represented Sweden on the organising committee of the 1932 Los Angeles Olympic Games.

Arie van Donkelaar.	Dutch.	K.	An official of a tugboat company returning from business meeting. His boss had decided not to return that day.
Doctor Kurt Hoene.	Polish.	K.	A couple believed to be returning from honeymoon.
Mme Ursula Hoene.	Polish.	K.	
Rear-Admiral Saloman Arvid Achates Lindman.	Swedish.	K.	Former Swedish Prime Minister; returning to Sweden after representing his country at a Masonic bi-centenary conference in Edinburgh.
Miss Marie Lipsey.	South African.	K.	On a business trip for a Johannesburg store.
Frederick G.O. Mergell.	German.	K.	Lived in Hamburg.
Baron Gottfried Maria Johannes Oskar Alfred Ernest Nikolaus von Mayern-Hohenberg.	Austrian.	K.	To Berlin to meet his ex-wife, the film star Luli Deste.
Sydney Pearson.	British.	K.	On business trip for The Phosphor Bronze Company Ltd. His home was just one mile from the crash site.
Leonard Stemp.	British.	K.	Worked for the British National Institute for the Blind and was developing 'talking books' for the blind. Travelling to Berlin in connection with his work.
Carl Magnus Wegelius.	Finnish.	K.	A commercial representative of a Finnish department store, returning to Finland.
Walter Schubach.	German.	Seriously Injured, survived.	Travelling to Berlin.

Appendix 2

Lombard

List of crew and passengers, all of whom died aboard Carole Lombard's Transcontinental & Western Air (TWA) DC-3, NC-1946, Flight 3 from Las Vegas to Burbank. 16 January 1942.

Crew
Wayne C. Williams, Captain.
Morgan A. Gillette, Co-pilot.
Alice F. Gett, Flight attendant.

Passengers (by seat numbers)
1. Sergeant A.M. Belejchak. US Army Air Corps.
2. Corporal M.B. Affrime.
3. Second Lieutenant K.T. Donaghue.
4. Second Lieutenant Hal E. Browne.
5. Second Lieutenant James B. Barham.
6. Second Lieutenant Stuart L. Swenson.
7. Private Nicholas Varsamino.
8. Carole Lombard, Hollywood star, going home after a War Bond sales tour.
9. Elizabeth K. Peters, Lombard's mother, who accompanied her daughter on the sales tour.
10. Second Lieutenant 2/Lt Charles D. Nelson.
11. Otto Winkler, Clark Gable's PR agent, who accompanied Lombard on her tour.
12. Sergeant Robert Nygren.
14. Second Lieutenant 2/Lt Frederick J. Dittman.
15. First Lieutenant Robert E. Crouch.
16. Staff Sergeant Edgar A. Nygren.
17. Lois Hamilton, an Army Air Corps wife, travelling to meet her husband.
18. Staff Sergeant David C. Tilghman.
19. Sergeant Frederick P. Cook.
20. Private Martin W. Tollcamp.

Note: The number 13 is not applied to any airliner seat.

Appendix 3

Busby's Babes

Passengers aboard BEA Ambassador, G–ALZU, on 6 February 1958.

Manchester United Players
Geoff Bent, died.
Johnny Berry, injured, never played again.
Jackie Blanchflower, injured, never played again.
Roger Byrne, died.
Bobby Charlton, injured, recovered to play again for United and England.
Eddie Colman, died.
Duncan Edwards, died in hospital of his injuries two weeks later.
Bill Foulkes, unhurt, played again for United.
Harry Gregg, thought to be unhurt but later diagnosed with fracture to skull, played again for United.
Mark Jones, died.
Kenny Morgans, injured, played again.
David Pegg, died.
Albert Scanlon, injured, played again for United.
Tommy Taylor, died.
Dennis Viollet, injured, played again for United.
Liam 'Billy' Whelan, died.
Ray Wood, injured, rarely played again.

Manchester United Staff
Matt Busby, injured, in hospital for two months.
Walter Crickmer, died.
Tom Curry, died.
Bert Whalley, died.

Journalists
Alf Clarke, died.
Don Davies, died.
Ted Ellyard, unhurt.
George Follows, died.

FATAL FLIGHTS OF THE RICH AND FAMOUS

Peter Howard, unhurt.
Tom Jackson, died.
Archie Ledbrooke, died.
Henry Rose, died.
Frank Swift, died
Frank Taylor, injured.
Eric Thompson, died.

Other Passengers
Verena Lukic, unhurt.
Vesna Lukic, unhurt.
Bela Miklos, died.
Eleanor Miklos, injured.
Willie Satinoff, died.
Nebosja Tomasevic, injured.

Crew
Margaret Bellis, unhurt.
Tom Cable, died.
Rosemary Cheverton, unhurt.
Captain Kenneth Rayment, died of his injuries in hospital three weeks later.
George Rodgers, unhurt.
Captain James Thain, unhurt.

Occupants of House
Unidentified mother and three children, unhurt.

Appendix 4

Hammarskjöld

Passengers and crew of Douglas DC-6B, SE-BDY, on charter from Swedish Transair Company to ONUC Military Operations, which crashed near Ndola, 18 September 1961. All sixteen souls on board died.

Crew (all Swedish nationality)
Pilot in command: Captain Per-Erik Bo Hallonquist.
First Officer (second pilot): Lars Olaf Litton.
Reserve captain: Captain Nils-Eric Aahréus.
Flight engineer: Nils Göran Wilhelmsson.
Radio operator: Carl Eric Gabriel Rosén.
Assistant purser: Harald Noork.

Passengers
Dag Hjalmar Agne Carl Hammarskjöld (Sweden), Secretary-General of the United Nations.
Heinrich A. Wieschhof (USA), Director & Deputy to the Under-Secretary of UN Political and Security Council Affairs.
Vladimir Fabry (USA), Special Counsellor to the Officer-in-Charge of UN Operations in The Congo.
William Ranollo (USA), Personal Aide to the UN Secretary General.
Alice Lalande (Canada), Secretary to the Officer-in-Charge of UN Operations in The Congo.
Sergeant Harold M. Julien (USA) Acting Chief Security Officer ONUC. Having survived the crash, he died of his injuries in hospital six days later.
Sergeant Serge L. Barrau (Haiti), Security Officer, ONUC.
Sergeant Francis Eivers (Ireland), Investigator, ONUC.
Warrant Officer Stig Olof Hjelte (Sweden), 11th Infantry Battalion, Swedish Army, Security Guard.
Private Per Edvald Persson (Sweden), 11th Infantry Battalion, Swedish Army, Security Guard.

Appendix 5

Thorburn

List of crew and passengers, all of whom died aboard Iberia Airways Sud-Aviation SE-210 Caravelle 10R, EC-BDD, *Jesus Guridi*, Flight IB-062 on 4 November 1967.

Crew
Hernando Maura Pieres, Captain.
José Moreno Humada, Co-pilot.
José Maria Alvarez Menendez, Flight Engineer.
Araceli Cedillo Casanova, Stewardess.
Rosario Gonzalez Fraile, Stewardess.
Hermenegildo Garcia Julia, Steward.
Manuel Leandro Barroso, Steward.

Passengers (in alphabetical order)
Mr Archibald Thomas Thain Alexander of Camberley, Surrey.
Mr William James Barrett, Brighton; fishmonger.
Mrs Constance Barrett; wife of above.
Mrs Susan Rosemary Bell.
Mr Bethune, Australia.
Mr Donald Campbell, Hungerford, Berks; MD of Campbell Aircraft Company.
Mrs Lilian (Nan) Campbell; wife of the above.
Mr John Clarkson, Coventry; Industrialist and Vice-President of Coventry City Football Club.
Mrs Rosemary Clarkson; wife of the above.
Miss Paula Clarkson (4); daughter of the above.
Miss Katie Clarkson (3); daughter of the above.
Mr J. H. Cotton, Lancashire.
Mr Neill Creagh-Chapman, Hans Place, London.
Mrs Patricia Creagh-Chapman; wife of the above.
Mrs Goldman.
Mr Charles Douglas Griffiths, Wick, near Bristol; deputy coroner for Bristol & South Gloucestershire.
Mrs Viola Maclean Griffiths; wife of the above.
Mrs Edith King, Ottershaw, Surrey.

THORBURN

Miss Elizabeth Luff; Guernsey; nanny to the Clarkson children.
Mr Arthur V. Madden.
Mr Mandell, USA.
Senor Josefa Moya Martin, Spain.
Miss Anita June Peer; a BOAC air stewardess.
Mrs Golda Peer, Manor Park, London; mother of the above.
Mr Perkins, USA.
Mrs Barry Raickes, Mount Street, London.
Mr Simon Nicholas Crane Richardson, Hertfordshire.
Senor Francisco Herrera Santos, Spain.
Mr Barry Markham Thomas, Hove, Sussex.
Miss June Thorburn (Mrs June Smith-Petersen), East Heath Road, London; actress.

Appendix 6

Lucas

Names of the 63 fatalities on BEA Vanguard, G-APEC, crashed in Belgium on 2 October 1971.

Crew
Probert, Captain Edward T.
Davies, First Officer John Michael.
Barnes, First Officer Bevil J.S.
Partridge, Captain Guy; supernumerary pilot for route experience.
McCarthy, William, Chief steward.
Forrow, Anthony, Steward.
Fetch, Hilary, Stewardess.
Johnson, Marylyn Joyce, Stewardess.

Passengers
British nationality:

Adams, Jack Hawthorn.
Billington, Deborah. Fiancee of Colin Rolfe (see below).
Blamey, Miss J.
Boyd, Ronald.
Boyd, Mrs Barbara Mary.
Breden, Mrs Maria Franziska.
Collins, John.
Collins, Mrs Nellie.
Collins, Miss Anne (18)
Collins, Anthony, teenage twin.
Collins, Geoffrey, teenage twin.
Crawford, Miss H.
King, Miss Margaret.
Lucas, Otto.
Manton, Miss Madeline Anne.
Marsh, John Bryan.
Matheson, Gavin Niall.

Odlin, Clifford.
Odlin, Mrs Emily.
Poulton, Mrs Raymonde Catherine Adelaide Wilhelmina.
Rippon, George Henry MBE.
Rolfe, Colin F.
Seton, George Werner.
Starkey, Mr G.S.
Starkey, Mr A. BEA staff.
Starkey, Mrs Margaret. BEA staff.
Sutherland, Ian Duncan.
Varley, Mr James Simon. BEA staff.
Varley, Mrs Dale. BEA staff.
Wakeman, Edward Henry. BEA staff; station engineer.
Walker, Miss Elizabeth Helen. Fiancee of Franz Zimmert (see below).

American nationality
Baerga, Miss Belen Colon.
Castro-Fernandez, Miss Georgina.
Castro-Fernandez, Miss H.
Kessell, Benjamin.
Kessell, Mrs Phyllis Barbara.
Lindberg, Stanley.
Lindberg, Mrs.
Moser, Mrs Joan Lindsley.
Wincor, Mrs Margaret Reber.

Austrian nationality
Aigner, Miss.
am-Tinkhof, Miss Anna Mair.
Braun, Dr H.
Marcic, Professor Rene.
Marcic, Mrs Blanka.
Moser, Mr R.
Safner, Miss Maria.
Thiem, Miss Beatrix.
Zimmert, Franz.

Japanese nationality
Ohira, Mr Tadashi.
Ohira, Mrs Toshiko,

Ohira, Keisuke. Son of the above.
Ohira, Mina. Daughter of the above.

Unknown nationality
Hullinger, Miss Sandra.
Wassermann, Mr A.

Endnotes

Chapter 1

1. Using: bankofengland.co.uk/monetary-policy/inflation/inflation-calculator.

Chapter 2

1. Leman, Jennifer, *Popular Mechanics*, 7 March 2022. Popularmechanics.com/science/
environment/a32496561/why-magnetic-north-pole-moving.

Chapter 3

1. EON Productions, 1967.

Chapter 4

1. Book title:*20 Hrs. 40 min. Our flight in the Friendship. The American Girl, first across the
Atlantic by air, tells her story*. G.P. Putnam's Sons, New York, 1928.
2. Using US Bureau of Labor Statistics: bls.gov/data/inflation_calculator.
3. Lutz, Amy, *Amelia Earhart; Myth and Mystery*, Thesis 365, 2020. Irl.umsl.edu/thesis/365,
pages 11-12.

Chapter 7

1. Goss, Chris, *Bloody Biscay*, Crecy Publishing Ltd, Manchester, 2001, pages 72-73.
2. International Pictures: 1939.
3. *London Gazette*, March 1920, published by The Stationery Office, under superintendence
of His Majesty's Stationery Office, part of The National Archives.
4. Warner Bros, 1936, USA.
5. General Film Distributors in UK; MGM in USA; 1938.
6. Selznick International; 1939.
7. MGM, 1931.
8. Anglo-American Film Corporation, 1941.
9. General Film Distributors in UK; Columbia Pictures in US, 1941.

Chapter 8

1. *Variety* Magazine, 6 January 1943, page 58.

Chapter 10

1. UK Parliament *Hansard*, Vol 753, cc 1402-12, Mr W.R. Straubenzee, 9 November 1967.
2. UK Parliament *Hansard*, Vol 784, cc 1234-8, 10 June 1969.

Chapter 11

1. US Bureau of Labor Statistics.data.bls.gov.

Chapter 12

1. Frielingsdorf, 2016, page 148.
2. Hammarberg report, 15 August 2013, pages 22-23.

Chapter 14

1. Leonev interview; rt.com/news/gagarin-death-truth-revealed-674.

Chapter 17

1. Nyberg, Chapter 6, pages 85-93.

Chapter 19

1. Heppenheimer, 1999, pages 193-194.

Chapter 20

1. Crash site in South Dakota is in US Central Time Zone and this equates to 12.12:26 Central Daylight Time/CDT.

Chapter 21

1. Data and information from the King Commission Interim Report, 11 August 2000, ISBN 0/62026433/0.

Chapter 22

1. Pprune.org post on 28 June 2013 and www.newser.com/story/161023/pilot-of-missing-missoni-plane-had-expired-licence.
2. Telephone interview by Associated Press correspondent, Wednesday 9 January 2013 was reported by ABC News online, 10 January 2013.
3. From online *Vogue* magazine article by Isabel Edwards-Brown, 14 February 2013. vogue.com.au/fashion/news.
4. www.bresciatoday.it/cronaca/los-roques-foresti, 20 August 2024.

Chapter 23

1. From an article in its 'Barstool Sports: The Corp,' section, by Jenna Amatulli, 22 January 2020.
2. Article: LATimes.com: Daily Pilot article by Luke Money, Gustavo Arellano and Hillary Davis; 29 January 2020.
3. NTSB Accident report NTSB/AAR-21/01, PB2021-100900, dated 9 February 2021.

Bibliography

Chapter 1

Brett, R. Dallas, *History of British Aviation, Vol 2*, (The Aviation Book Club, London,1928).

Brown, Arthur Whitten, *Flying the Atlantic in Sixteen Hours*, (Frederick A. Stokes Co., New York, 1920).

Lynch, Brendon, *Yesterday We Were In America*, (Haynes, London, 2009).

Wallace, Graham, *The Flight of Alcock & Brown*, (Putnam, London, 1955).

Daily Herald, 17/6/1919, page 5.

Globe, The, 16/6/1919, page 8.

lancashirelife.co.uk.

Le Petit Parisien, (Paris Edition), 20 December 1919, page 1: 'Alcock s'est tué.'

projectgutenberg.org.

scienceandindustrymuseum.org.uk.

Sheffield Evening Telegraph, 17/6/1919, page 6.

Chapter 2

Amundsen, Roald, *My Life as an Explorer*, (Amberley Publishing, Stroud, 2014).

Cameron, Garth, *Umberto Nobile and the arctic search for the airship Italia*, (Fonthill Media, Stroud, 2017).

Kristensen, Monica, *Amundsen's Siste Reise*, (Forlaget Press, Oslo, 2017).

coolantarctica.com.

Fram – The Polar Exploration Museum, Oslo, Norway. (frammuseum.no).

Italia, the airship crash chronicle; Italia.tass.com, 2018.

Archive of National Library of Norway; Nasjonalbiblioteket, Oslo. www.nb.no.

Chapter 3

Brooks, Peter W., *Cierva Autogyro; the Development of Rotary-wing Flight*, (Smithsonian Press, USA, 1988).

Haine, Edgar A., *Disasters in The Air*, (Cornwall Books, New York, 2000).

Jackson, A.J., *British Civil Aircraft since 1919, vols 1-3*, (Putnam, London, 1973).

Phipp, Mike, *Wessex Aviation Industry*, (Amberley Publishing, Stroud, 2011).

aviacrash.nl.

aviastar.org/cierva helicopters and autogyros.

Croydon Times & Surrey County Mail.

Lewis, Jeff, *Autogyro History and Theory*, essay, USA, 1996.

rcrawsey.co.uk.

resources.huygens.knew.nl.

Chapter 4

Chapman, Sally Putnam, *Whistled Like A Bird. The untold story of Dorothy Putnam, George Putnam and Amelia Earhart*, (Warner Books Inc., New York, 1997).

Long, Elgen M., & Long, Marie K., *Amelia Earhart: The Mystery Solved*, (Simon & Schuster, New York, 1999).

Lutz, Amy, *Amelia Earhart: Myth and Memory*, (Thesis 365, University of Missouri, St Louis, 24 July 2020).

Miller-Cribbs, J. & Mains, J., *Amelia Earhart (1897-1937): Social worker, women's advocate, world famous American aviation pioneer. Social Welfare History Project, 2011.* Retrieved 3 May 2021, from http://socialwelfare.library.vcu.edu/people/earhart-amelia/.

Rich, Doris L., *Amelia Earhart, A Biography*, (Smithsonian Books, Washington DC, 1989).

Winters, Kathleen C., *Amelia Earhart: The Turbulent Life of an American Icon*, (Palgrave Macmillan, New York, 2010).

e-archives/lib.purdue.edu.

Chapter 5

King, Alison, *Golden Wings*, (C. Arthur Pearson Ltd., London, 1956).

Lewis, Peter, *British Racing & Record-breaking Aircraft*, (Putnam, London, 1970).

Lomax, Judy, *Women of the Air*, (John Murray, London, 1986).

Smith, Constance Babington, *Amy Johnson*, (Collins, London 1967).

Hull History Centre; *The Amy Johnson Letters.* hullhistorycentre.org.uk.

Amy Johnson Flying Log book online: eastridingmuseums.co.uk.

Chapter 6

Kiriakon, Olympia, *Becoming Carole Lombard: Stardom, Comedy and Legacy*, (Bloomsbury Academic, New York & London, 2020).

Matzen, Robert, *Fireball: Carole Lombard and the mystery of Flight 3*, (Goodknight Books, Pittsburgh, USA, 2017).

US Civil Aeronautics Board: *Report 119-42/SA-58*; issued 16 July 1942, books.google.co.uk, US House of Representatives, *Select Committee on Air Accidents: Report No. 2323*, pages 33-35, 1942.

Sky Arts TV, *Discovering: Carole Lombard*, Season 8; Episode 12: Wed 2 August 2023.

BIBLIOGRAPHY

Greeley (Colorado) Daily Tribune, Sat 17 January 1942, page 1.
carole-and-co.livejournal.com.
crystalkalyana.wordpress.com (*In the good old days of Classic Hollywood*).
indianahistory.org.
lostflights.com.
vp19.livejournal.com.

Chapter 7

Colvin, Ian, *Flight 777; The Mystery of Leslie Howard*, (Pen & Sword, Barnsley, 2013).

Eforgen, Estel, *Leslie Howard: The Lost Actor*, (Vallentine Mitchell, London & USA, 2013).

Goss, Chris, *Bloody Biscay, The history of V Gruppe/KG40*, (Crecy, Manchester, 2001).

Howard, Ronald, *In Search of My Father; A Portrait of Leslie Howard*, (St Martin's Press, London, 1984).

Chapter 8

Neil, Wing Commander Tom, DFC*, *The Silver Spitfire*, (Phoenix, London, 2014), page 235.

Spragg, Dennis M., *Glenn Miller Declassified*, (Potomac Books-University of Nebraska Press, USA, 2017).

Spragg, Dennis M., *75th Anniversary commemoration of the film: Sun Valley Serenade;* from

Way, Chris, *Glenn Miller in Britain, Then and Now*, (After the Battle, London, 1996).
Big Band Library.com.
Daily Express.
Glenn Miller Archive, American Music Research Center, University of Colorado, Boulder, USA; 2016.
Harpenden & District Local History Society. Harpenden-history.org.uk
IMDb.com.
International Group for Historical Aircraft Recovery, The. TIGHAR.org.
skeptoid.com.
Variety Magazine, published weekly by Variety Publishing Company Inc, New York USA.
Guidelondon.org.

Chapter 9

Hancock, Terry, *Directory of Britain's Military Aircraft, Vols 1 & 2*, (The History Press, Stroud, 2008 & 2010).
Jackson, A. J., *Avro Aircraft since 1908*, (Putnam, London, 1990).

Jackson, A. J., *British Civil Aircraft since 1909, Vol 1*, (Putnam, London, 1973).

National Archives, Kew, London: Accident report on G-AGSU by UK Chief Inspector of Accidents, Ref BT217/1926.

Oxford Dictionary of National Biography, *Chadwick, Roy*; 1959 and 2008 editions.

Penrose, Harald, *Architect of Wings, Biography of Roy Chadwick, designer of the Lancaster Bomber*, (Airlife Publishing, London, 1985), via archive.org.

Penrose, Harald, *British Aviation: The Great War and Armistice*, (Putnam, London, 1969).

Penrose, Harald, *British Aviation: Widening Horizons*, (HMSO, London, 1979).

Penrose, Harald, *British Aviation: Ominous Skies*, (HMSO, London, 1980).

Poolman, Kenneth, *Zeppelins Over England*, (Evans Brothers Ltd, London, 1960).

Chapter 10

Morrin, Stephen R., *The Munich Air Disaster – The true story behind the fatal 1958 crash*, (Gill & Macmillan, London, 2007).

Roberts, John, *The Team That Wouldn't Die: The story of the Busby Babes*, (Aurum Press, London, 2008).

Hansard, UK Parliament. (Public sector information licensed under the Open Government Licence v3). (Parliamentary information licensed under the Open Parliament Licence v.3).

Chapter 11

Bie, Michael, *Wisconsin Myths & Legends*, (Rowman & Littlefield, Lanham USA, 2022).

Everitt, Rich, *Falling Stars: Air crashes that filled Rock and Roll Heaven*, (Harbor House, Augusta USA, 2004).

Kocis, Desiree, *Mysteries of Flight: The day the music died*, (Pilot and Piloting Magazine, USA).

Leigh, Spencer, *Buddy Holly: Learning the Game*, (McNidder & Grace, London, 2019).

Norman, Philip, *Buddy: The definitive biography of Buddy Holly*, (Macmillan, London, 1996).

Pilotandpilotingmag.com. 4 February 2020.

US Civil Aeronautics Board (CAB) Report; File No. 2-0001.

Chuckberry.com: Winter Dance Party Tour 1959.

Daily Mail, 25 February 2019, Answers to Correspondents, page 54.

Tribute to Buddy Holly by Buddy Walker: buddywalker.co.uk.

Chapter 12

Chesterman, Simon (ed.), *Secretary or General? The UN Secretary-General in World Politics*, (Cambridge University Press, Cambridge, 2007), page 6.

BIBLIOGRAPHY

Frielingsdorf, Per-Axel, *Machiavelli of Peace: Dag Hammarskjöld and the Political Role of the Secretary-General of the United Nations.* PhD Thesis for Dept of International Relations of the London School of Economics & Political Science; London, March 2016.

Louis, William Roger, *Harold Macmillan and the Middle East Crisis of 1958*, University of Texas. Paper read at British Academy 22 October 1996. Proceedings 94, pages 207-228.

Hammarberg, Sven E., *Accident Investigator's report to the Hammarskjöld Commission of Jurists set up to enquire into the death of Dag Hammarskjöld*, Kristianstad, Sweden, 15 August 2013.

Krasno, Jean, *Interview with Sture Linnér*, Yale University Oral History Project on UN. New York, November 1990, pages 34-36.

Othman, Mohamed Chande, *Investigation into the death of Dag Hammarskjöld*; A report to UN Security Council dated 12 September 2019; ref: A/73/973; documents-dds-ny.un.org, N1924622.pdf.

Rösiö, Bengt, *The Ndola Disaster. Revised version*, for Swedish Ministry for Foreign Affairs November 1992 – February 1993.

Urquhart, Brian, *Hammarskjöld*, (Knopf, New York, 1972).

Report of the UN Commission of Investigation into the circumstances of the death of the UNSG. Ref A/5069 dated 24 April 1962.

Report of Federation of Rhodesia & Nyasaland Commission into accident to SE-BDY, chaired by Sir John Claydon, Chief Justice of Federation; Feb 1962.

Report of Director of Civil Aviation, Federation of Rhodesia & Nyasaland, in IACO Circular 69-AN/61, AR703, Feb 1962, p183.

un.org: e.g. UN Observation Group in the Lebanon (UNOGIL) 1958.

Chapter 13

Board of Trade report CAP 343, published by HMSO, via Flight International archive.

British Newspaper Archive Online.

Broadwayworld.com/westend.

Fernhurst Society, The. fernhurstsociety.org.uk/caravelle; 2010.

Filmsofthefifties.com.

IMDb.com.

John-clarke.co.uk. Historian of Brookwood cemetery.

Ministry of Civil Aviation, air accident report, AVIA101/555.

Open University, Open.ac.uk/arts/history-from-police-archives.

Radio Times archive.

Theaticalia.com: Plays, people, places.

Times, The (London).

Chapter 14

Jenks, Andrew l., *The Cosmonaut who couldn't stop smiling; The life & legend of Yuri Gagarin*, (Northern Illinois University Press, USA, 2014).

Singh, Gurbir, *Yuri Gagarin in London & Manchester*, (Astrotalk Publishers, Manchester, 2011).

Walker, Stephen, *Beyond; The Astonishing Story of the First Human to Leave our Planet and Journey into Space*, (William Collins, London, 2021).

Astronautix.com.

Anatoly Zak: Russianspaceweb.com.

Chapter 15

Stanton, Mike, *Unbeaten: The Triumphs and Tragedies of Rocky Marciano*, (Henry Holt & Co, New York, 2018) and (Macmillan, London, 2018).

Sullivan, Russell, *Rocky Marciano, The Rock of his Times*, (University of Illinois Press, USA, 2002).

Des Moines Register; article: 2 September 1969, page 3.

Cedar Rapids Gazette; article: 1 September 1969.

boxrec.com.

WBO.com.

Chapter 16

Murphy, Audie, *To Hell and Back*, (Henry Holt & Co, New York, 1949).

audiemurphy.com (Audie Murphy Research Foundation and official Audie Murphy website).

publish.iupress.indiana.edu/shooting-stars: *Audie Murphy: Kid with a Gun*; a paper by Don B. Graham; undated.

JG Photographics.com.

King Features Syndicate and The Farmersville Times/C&S Media, transcripts of interviews with Audie Murphy, circa August 1945.

Ryan,Carolyn Price, Army Nurse Corps; 12 February 1973, her recollections of Audie Murphy, audiemurphy.com.

Rider to the recollections of Carolyn Ryan, by Spec McClure, 12 Oct 1973.

charliecompany.org; articles by Vicki Ellis Griffis, written for *Celeste Tribune*. Date unknown.

Chapter 17

Nyburg, Dr Anna, *The Clothes On Our Backs; How refugees from Nazism revitalised the British fashion trade*, (Valentine Mitchell, London, 2020).

Parkin, Sophie, *The Colony Room Club, 1948-2008, A History of Bohemian Soho*, (Palmtree Publishers, London, 2012).

British Newspaper Archive online.

BIBLIOGRAPHY

Times, The (London), list of passengers and crew.

Sydney Morning Herald.

University of Munster: documents relating to pleas for help from the Pope. Also: Archivio Apostolico Vaticano, Segretaria di Stato, Commissione Soccorsi 302, fasc.,9, fol.,13rv.

US Holocaust Memorial Museum database. Uni-munster.de/FB2/aph/bittschreiben/lucas.html

www.joodsmonument.nl.

National Archives (UK): Home Office records: Naturalisation documents: e.g. HO 334/843/66363.

National Archives (UK): Home Office records: WW2 Internee Tribunal documents: e.g. HO396; accessed via findmypast.co.uk.

Bide, Bethan; *Austerity Fashion, 1945-1951*. PhD Thesis; June 2017, via core.ac.uk.

Borthwick Institute for Archives, University of York.

Museum of London archive.

Probate records at gov.uk.

Nyburg, Dr Anna, Imperial College, London, interview with Rolf Andresen, August 2017.

UK Air Accident Investigation Branch Reports: No. 15/1972, published December 1972; via gov.uk.

London School of Economics; transcript of interview of Philip Somerville by Nikki Cheetham, dated 15 March 1988; part of LSE's *Out of the Doll's House* collection, ref 8ODH, in LSE Library.

British Library; National Life Stories ref: C1046/06; *An Oral History of Fashion*; transcript of interviews of Frederick Fox by Linda Sandino, recorded between 15 Sept 2003 and 14 June 2004.

Chapter 18

Tipler, John, *Graham Hill: Master of Motorsport*, (Breedon Book Publishers Ltd, Stoke, 2002).

www.atlasf1.com.

www.blackebushairport.proboards.com.

Chapter 19

Boyd, Robert J., *SAC's Fighter Planes and their Operations*, (HQ SAC, USA,1988).

Heppenheimer, T.A., *The Space Shuttle Decision*, (NASA History Series, Washington DC, 1999).

Merlin, Peter W., *Unlimited Horizons: Design & development of the U-2*, (NASA, USA, 2015).

Polmar, Norman, *Spyplane: The U-2 History Declassified*, (MBI Publishing, USA, 2001).

Powers, Francis Gary, with Gentry, Curt, *Operation Overflight: A Memoir of the U-2 Incident*, (Holt, Rinehart & Winston, USA, 1970, and Potomac Books, 2004).

Powers, Francis Gary Jnr and Dunnavant, Keith, *Spy Pilot, Francis Gary Powers, the U2 incident and a controversial cold war legacy,* (Prometheus Books, New York, 2019).
adst.org; Toumanoff interview by William D. Morgan; ADST, 1999.
alternatewars.com.
brownpundits.com, archived article by Hamid Hussein, 29 Oct 2010.
cia.gov/readingroom/historical-collections
earlytelevision.org/helicopters.
eisenhowerlibrary.gov.
lamorguefiles.blogspot.com.
laradio.com.
Report of *Hearing before US Senate Committee on Armed Forces, 6 March 1962,* (US Government Printing Office, Washington, 1962).
Smithsonian Online Virtual Archives; sova.si.edu.

Chapter 20

Negroni, Christine, *The Crash Detectives*, (Atlantic Books, London, 2016).
Robbins, Kevin, *The Last Stand of Payne Stewart*, (Hachette Books, New York, 2019).
Stewart, Tracey with Abraham, Ken, *Payne Stewart: The Authorised Biography*, (Broadman & Holman, Nashville, 2000).
NTSB Aircraft Accident Brief, Accident Number DCA00MA005; 28 November 2000; via skybrary/aero/bookshelf/books/315.pdf.
On Guard, The; newspaper of the Army & Air National Guard, Vol XXVIV, No.1, Nov 1999.

Chapter 21

King, Garth, *The Hansie Cronje Story, Authorised Biography,* (Global Creative Studios, Brakenfell SA, 2005).
Redford, Brian, *Caught Out; Shocking Revelations of Corruption in International Cricket*, (John Blake Publishing Ltd, London, 2012).
Wisden Cricketer's Almanac, (Bloomsbury Publishing, London, 2003).
Cricketcountry.com, article by Arunabha Sengupta, 2016.
espncricinfo.com
Guardian, The.
IOL.co.za/news/south africa/cronje inquest; 7–8 August 2006.
King Commission Interim Report; 11 August 2000. ISBN 0/620/26433/0.
South African Civil Aviation Authority, Air Accident Report, CA18/2/3/7510.

Chapter 22

Capella, Massimiliano, Boselli, Mario, *Missoni: The Great Italian Fashion*, (Rizzoli International Publications, New York, 2019).

BIBLIOGRAPHY

Fogg, Marnie, *Vintage Fashion Knitware; Collecting and wearing Designer Classics*,(Carlton Books, London, 2010), page 166.

ABC News online, 10 January 2013.

Aerialvisuals.ca.

Bresciatoday.it, Chronicle Pralboino, 20 August 2024.

BBC.co.uk/news/world-latin-america-24624548, 22 October 2013.

Elle online, article by Lauren Levinson, 17 October 2013.

Guardian, The, (UK) online, article by Tyler McCall, 17 October 2013.

Hydro-international.com/content/news, 1 July 2013.

Il Giornale, article by Fausto Biloslavo, 8 January 2013.

Junta Investigodora de Accidentes de Aviación Civil (JIAAC) report into accident to Let410, YV2081, 2008. Los Angeles Times, article by Booth Moore, 27 June 2013.

Marinelink.com, 10 July 2013.

San Diego Tribune.com, 7 January 2013.

Vogue.com.au/fashion/news, Isabel Edwards-Brown, 14 February 2013.

Vogue.co.uk, article by Ella Alexander, 27 June 2013.

Chapter 23

Christensen, Kim, Mazingo, Joe, Ormseth, Matthew, Vartabedian, Ralph, article: *Inside the last flight of Kobe Bryant*, Los Angeles Times/Journal Star, losangelestimes.com, 2020.

Huffpost.com.

National Air Transport Safety Board (USA), accident report: NTSB/AAR-21/01 PB2021-100900, 2 February 2021.

New York Times. newyorktimes.com

Names Index

Places Index

Dear Reader,

We hope you have enjoyed this book, but why not share your views on social media? You can also follow our pages to see more about our other products: facebook.com/penandswordbooks or follow us on X @penswordbooks

You can also view our products at www.pen-and-sword.co.uk (UK and ROW) or www.penandswordbooks.com (North America).

To keep up to date with our latest releases and online catalogues, please sign up to our newsletter at: www.pen-and-sword.co.uk/newsletter

If you would like a printed catalogue with our latest books, then please email: enquiries@pen-and-sword.co.uk or telephone: 01226 734555 (UK and ROW) or email: uspen-and-sword@casematepublishers.com or telephone: (610) 853-9131 (North America).

We respect your privacy and we will only use personal information to send you information about our products.

Thank you!